33 2/90 6

D0821756

912.09
L773a

34063

Lister
Antique maps and their cartographers

DATE DUE

GAYLORD 234   PRINTED IN U. S. A.

ANTIQUE MAPS AND THEIR CARTOGRAPHERS

By the same author
How to Identify Old Maps and Globes
William Blake
Edward Calvert
Decorative Cast Ironwork in Great Britain
Great Craftsmen
The Craftsman in Metal
Great Works of Craftsmanship

ANTIQUE MAPS AND
THEIR CARTOGRAPHERS

RAYMOND LISTER

LONDON
G. BELL AND SONS LTD

Published by G. Bell & Sons Ltd
York House, Portugal Street, London, W.C.2

SBN 7135 1586 4

Printed in Great Britain by
The Camelot Press Ltd., London and Southampton

FOR

BOB AND TILLY

WITH AFFECTION

Contents

The Plates

9

CHAPTER ONE

The Early History of the Map

THE craft of mapmaking is an ancient human accomplishment. Ancient man (and primitive people of our own time) knew at least its rudiments, even if their ideas of the shape and size of the Earth as a whole were usually mistaken. Sometimes, in a purely local sense, their maps were reliable. A section of coastline, a road, or the layout of a town or village, for instance, would be accurately drawn. But it is not surprising that their representations of the world, of a continent or even of a single country, were often wildly inaccurate. But before looking at the ancient history of cartography, let us for a moment look at the cartographic methods of some primitive peoples nearer our own time, thereby throwing some light on what might be termed the prehistory of cartography, and at the same time giving some indication of how the science came into being. We have no evidence that prehistoric man made his maps in the ways which I am about to describe, but perhaps we shall not be wholly mistaken in accepting the evidence of what more recent primitive people have done, as an indication of what might have happened in remote times.

The usual materials used by primitive people for their maps are wood and stone and, more rarely, skins and bone. The details are normally scratched or drawn on the surface or, in the case of stone and wood, chiselled or carved. Among some races, particularly the North American Indians, birch bark has been used. The Indians were expert cartographers and carried maps with them on their wanderings.

Eskimos, too, are good mapmakers and are alone among primitive people in depicting relief. Some of their maps are modelled in sand, with pieces of stone and wood used to mark elevations, villages and islands. Some are drawn on paper, with the hills shaded, as in more sophisticated maps. Most interesting of all are their wooden maps of the Greenland coastline, with the various features carved along the edge, giving a relief model of the area. Sometimes wood is saved by carving both edges of a plank, one edge a continuation of the other.   Islands are joined to the main plank by rods.

Some interesting sea charts were until comparatively recently made by the Marshall Islanders (Pacific Ocean).   Lengths of palm-fibre were tied together in a kind of grid, and sometimes shells, indicating islands, were placed at the intersections.   The lengths of fibre indicate the movement and direction of prevalent wave crests, and the points at which certain land features come into sight.   The Islanders' canoes sailed in groups under a leader who carried one of these charts.

The ancient Mexicans were efficient cartographers.   The Spanish conqueror of Mexico, Hernando Cortés, wrote in a report of 1520 to the Emperor Charles V how Montezuma, the ruler of Mexico, had the coastline charted for him and painted on cloth.   In a later report he spoke also of a map on cloth 'of the whole land, whereby I calculated that I could very well go over the great part of it'.

But to come to the mainstream of our subject.   The ancient Babylonians imagined the world to be a disc surrounded by water—'the Earthly ocean'—around the periphery of which were seven triangular islands.   One of these islands was supposed to be always in darkness, and it is thought that this might indicate some awareness of the long polar night.   Beyond the seven islands was the 'Heavenly ocean' in which the constellations circled.

The idea of the Earth as a sphere was first mooted by followers of Pythagoras the Ionian during the sixth century B.C., but it was a long time before it was generally accepted. Even during the Middle Ages not all geographers would have agreed with this idea.   The Greeks were very good geographers and a Greek astronomer, Eratosthenes (276–195 B.C.), worked out the circumference of the Earth to within fifty miles of the correct figure.   The first written reference to a map was made by the Greek historian Herodotus of Halicarnassus, though it was in the form of a sneer: 'For my part, I cannot but laugh when I see numbers of persons drawing maps of the world without having any reason to guide them; making, as they do, the ocean-stream to run all round the Earth, and the Earth itself to be an exact circle, as if described by a pair of compasses, with Europe and Asia just of the same size.'

This description is reminiscent of the Babylonian idea of the world as a disc surrounded by water.   Such Babylonian maps, moulded in clay, still exist.   The first Greek map, which was made by Anaximander of Miletus (610–546 B.C.), was probably drawn on the same lines.   Another map was made somewhat later by Aristagoras of Miletus.   It was of iron and was used by Aristagoras to induce the Spartans to plan a campaign against the Persians.   It was a Greek, Marinos of Tyre (*fl. circa* 100 B.C.), who first suggested the idea of projections on maps—that is, the reduction to a plane surface of the whole or part of the Earth's spherical surface.[1]

But there were other ideas of the Earth's form. A Greek manuscript of the sixth century B.C. described the Earth in anthropomorphic terms—head and face in the Peloponnesus, legs in the Hellespont, and so on. This was a forerunner by many centuries of William Blake's description of his mythical hero Albion in *Milton*:

> . . . his left foot near London
> Covers the shades of Tyburn: his instep from Windsor
> To Primrose Hill stretching to Highgate & Holloway.
> London is between his knees, its basements fourfold;
> His right foot stretches to the sea on Dover cliffs, his heel
> On Canterbury's ruins; his right hand covers lofty Wales,
> His left Scotland; his bosom girt with gold involves
> York, Edinburgh, Durham & Carlisle, & on the front
> Bath, Oxford, Cambridge, Norwich; his right elbow
> Leans on the Rocks of Erin's Land, Ireland, ancient nation.
> His head bends over London . . .

One of the most influential among ancient geographers was the astronomer, Klaudius Ptolemaios of Alexandria (A.D. 87–150), popularly known as Ptolemy. He wrote several works, the most famous being his *Geography*, which is in eight books, illustrated by maps. As an astronomer Ptolemy was a follower of Hipparchus, who founded scientific astronomy and discovered the precession of the equinoxes; as a geographer he was a follower of Marinos of Tyre. It is due to Ptolemy that we know of the work of these two men.

There are twenty-six regional maps and a *mappa mundi*[2] in Ptolemy's *Geography*. This is known to scholars as the 'A' recension. In addition to this there is a group of sixty-seven maps of smaller portions of the Earth's surface, which is known as the 'B' recension. The 'A' group was the one used for Latin manuscript editions of the *Geography* in the fifteenth century. The *mappa mundi* is drawn on a conic projection

---

[1] There are various methods by which projections may be made. See H. S. L. Winterbotham, *A Key to Maps* (London 1947) and A. R. Hinks, *Map Projections* (Cambridge).

[2] *Mappa mundi* = a world map. Derived from *mappa*, a napkin or towel, and *mundi*, of the world. The word 'map' is itself derived from this source. To the Romans, the word *mappa* denoted the games, and was derived from the fact that Nero signalled for the games to commence by throwing a napkin, on which he had wiped his hands, through a window. The word 'cartography', the study or making of maps, is derived from two Greek words χάριης, a leaf of papyrus or paper, and γραφειν, to write.

Ptolemy's *mappa mundi* covered only one hemisphere.

(Plate 1), but all the other maps are rectangular with right-angle intersections of parallels and meridians.

The earliest existing manuscript map based on Ptolemy is no earlier than the twelfth century, so it is uncertain if he was really such a great cartographer as his reputation indicates, or even whether he was a cartographer at all. It may merely be that the maps have been put to his credit because of his reputation as a geographer. We shall probably never know. But what is important is that he is at the centre of a certain tradition in cartography, and that certain maps and projections have been ascribed to him. For many centuries the work attributed to him was the accepted standard among Arab geographers, although until the fifteenth century he had little influence on western cartography.

Ptolemy's cartography was inaccurate according to modern standards, yet remarkable for the times in which it was conceived. The Indian Ocean is shown as a land-locked sea, the shape of India itself bears no resemblance to its true shape, and Ceylon, or Taprobana as he calls it, is shown as a huge island as big as Spain and France together. He exaggerated, too, the breadth of Europe and Asia, and thereby distorted many other areas. But he did show some awareness of the true layout of the world, and although he made guesses to fill in blanks, he was not so culpable in this respect as many later cartographers.

Other ancient peoples contributed to the development of cartography, but no contribution was as important as that of the Greeks. The Ancient Egyptians drew plans re-defining boundaries after the inundations of the Nile, and other plans showing the layout of mines and of buildings. But it was not until much later, when Egypt became hellenised, that she made any serious scientific contribution to mapmaking.

The Romans were practical map and plan makers, but their contribution to the theoretical side of the subject was far less than that of the Greeks. Few of their maps have survived. There are the 'Tabula Peutingeriana' (named after the sixteenth-century German collector, Konrad Peutinger) a road map; sketch maps of garrisons, and surveyors' maps, but little more.

One type of world map often used in Roman times was the T-O map—a map that is, that symbolically represents the world by means of a T within an O. The O represents the boundary of the world with the 'Earthly ocean', as conceived by the Babylonians, around it. Within this is the T, the upright of which represents the Mediterranean Sea, and the horizontal bar the meridian reaching from the River Nile to the River Don. The segments around the T represent the continents of Europe,

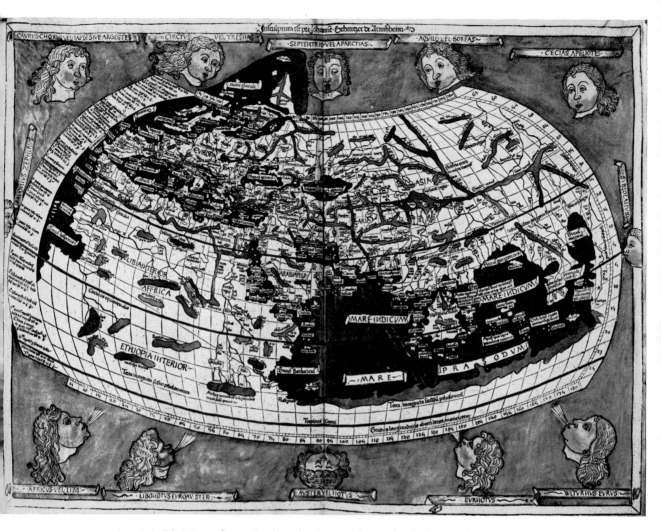

1. World Map, from Ptolemy's *Geography*, published at Ulm in 1482.
British Museum.

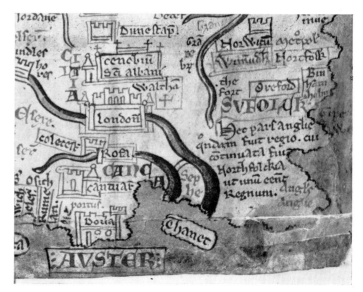

2. Detail from manuscript map of England *circa* 1250, by Matthew Paris. MS. Cotton Claud, D.VI.   *British Museum.*

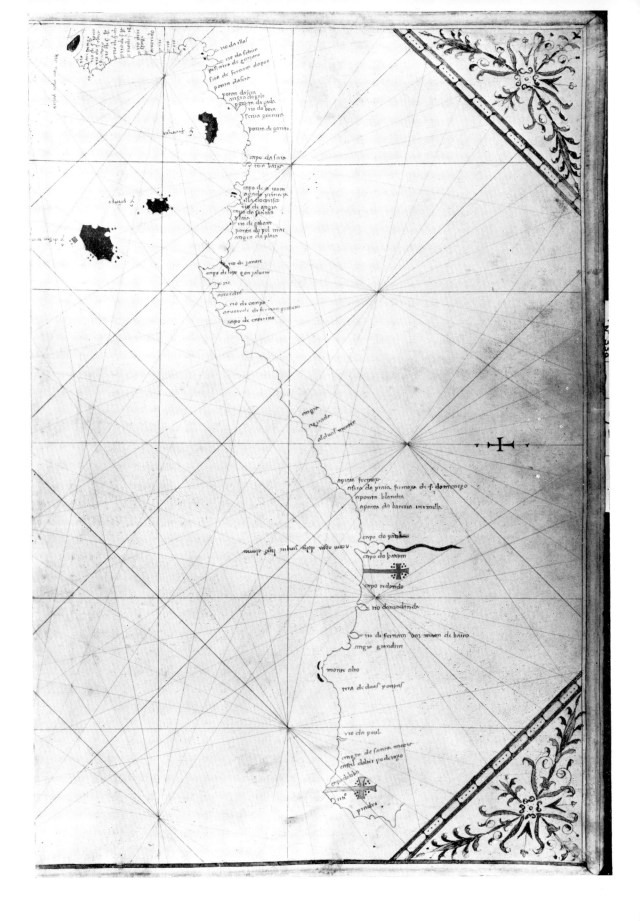

4. *Asiae XI Tabula* from Mercator's edition of Ptolemy. Cologne, 1578.
British Museum.

3 (*facing page*). Venetian chart of explora-
tion to the Congo by Diogo Cão, *Ginea
Portugalexe, circa* 1490. Egerton MS. 73.
British Museum.

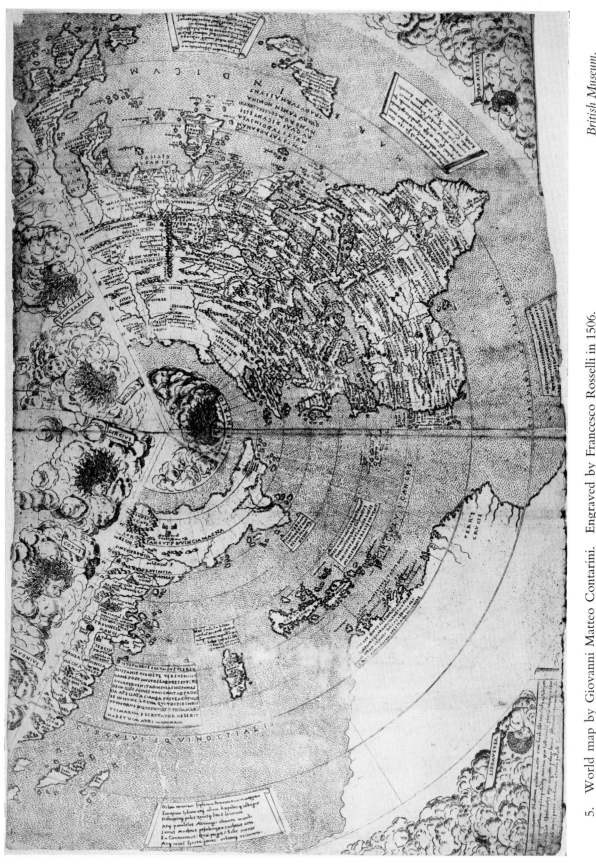

5. World map by Giovanni Matteo Contarini. Engraved by Francesco Rosselli in 1506.

British Museum.

6. *Angliae, Scotiae et Hiberniae* from Ortelius' *Theatrum Orbis Terrarum*, 1570.

British Museum.

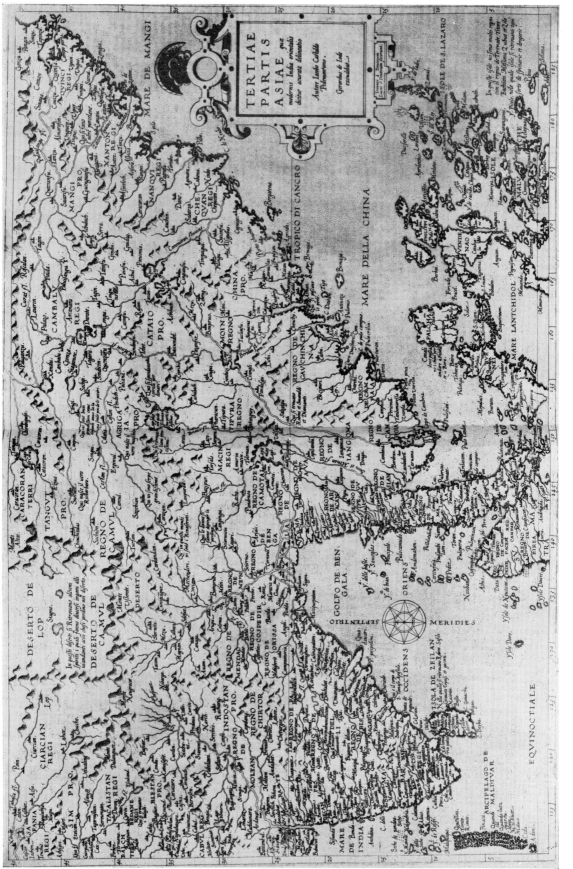

7. East Asia copied in 1578 by Gerard de Jode, from Gastaldi's map of 1561.   In de Jode's *Speculum Orbis Terrarum*.   *British Museum.*

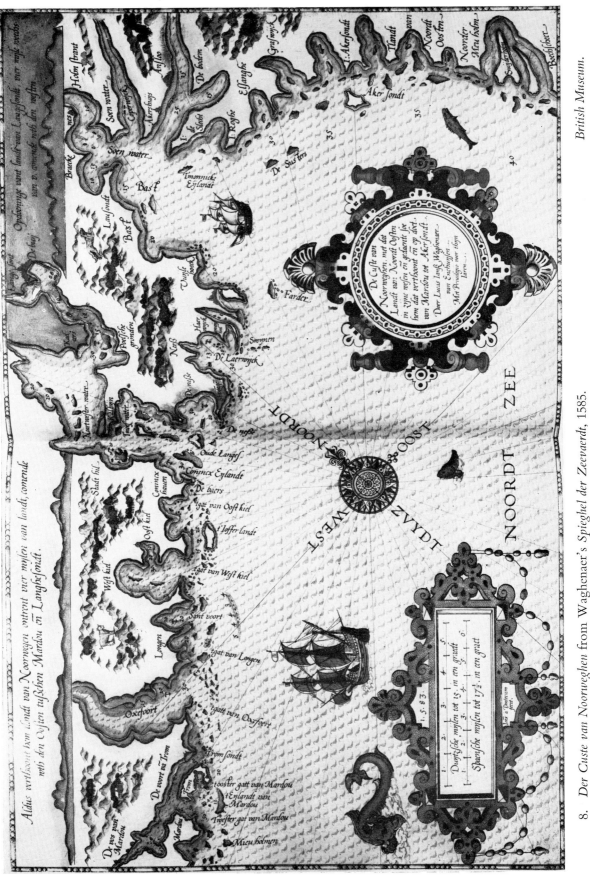

8. *Der Custe van Noorweghen* from Waghenaer's *Spieghel der Zeevaerdt*, 1585.

British Museum.

9. World map by Peter Apian, 1530.   *British Museum.*

Asia and Africa.   Variants of the T–O had a Y instead of a T within the circle, and a V within a square.   At least one map of the T–O type embodies the anthropomorphic idea described on page 15.   In this the T is formed by the figure of Christ crucified, with his head to the east (top), feet to the west (bottom) and his arms pointing to north (left) and south (right).   Jerusalem is at the centre.

T–O maps and their variants were popular in the Middle Ages and similar forms were widely used by Arab cartographers, although the Arabs' contributions to cartography were concerned more with the heavens than with the Earth.   None of these maps is of practical geographical significance; they are purely symbolic and diagrammatic.

The first printed T–O map, from a work by St. Isidore of Seville, printed in 1482

The same may be said of mediaeval cartography, in general, but decoratively it was often magnificent, as colourful and bright as illuminated manuscripts.   One magnificent mediaeval map is the famous *mappa mundi* in Hereford Cathedral.   It is circular, dates from about 1300, and is probably a close copy of a Roman original.   East is at the top, as in the T–O maps, and Jerusalem is shown as the centre of the world—a feature unlikely to have been found in the Roman original.   Its elaborate decorations include pictures of Paradise and of the Last Judgement.   Less care has been taken to represent the actualities of the earth's surface.   The British Isles, for instance, are squashed into a bay near to the perimeter of the circle and are grossly misshapen.   Scotland appears to be an island.

A somewhat more accurate representation of Great Britain is that made about 1250 by Matthew Paris, a Benedictine of St. Albans Abbey, who has been described as the first scientific cartographer in Britain.   His map is now in the British Museum[1] (Plate 2).   The map is small and bears only a rough resemblance to the shape of Great

[1] There are other maps of Great Britain by Matthew Paris, including a fragment in the Library of Corpus Christi College, Cambridge.

Britain, yet the beginnings of accuracy are present, and some topographical features are carefully portrayed. The general shape was probably deliberately distorted, so that the route from Doncaster to Dover could be accentuated—Dover being the point of embarkation for pilgrims to Rome and elsewhere. Special prominence is given to towns and cities with abbeys and monasteries—that is about nine-tenths of the total portrayed—and each is represented by a sketch of a religious building surmounted by a cross. Other towns are represented by battlemented buildings, and sometimes cross-surmounted and battlemented buildings are combined, as in the case of Canterbury. Each town is surrounded by a red line, which may have been meant to portray tiled roofs, which were by this time replacing the more easily combustible thatch. Although there are no roads, rivers are shown with great prominence, and are coloured blue.

About a century after the Paris map another well-known manuscript map of Great Britain was made. This is the Gough map of about 1360, called after the eighteenth-century topographer Richard Gough, who owned it and described it in his *British Topography* (1780). It is much closer than the Matthew Paris map to the true shape of this island. 'It is,' wrote Gough, 'drawn on two skins of vellum, in a style superior to any of the maps already described. The principal places are distinguished by churches with towers or spires; the rest by single houses. (These represent villages; and there is good reason for this . . . many of our villages have grown to be such from the house of a single considerable person, from whom they were named by adding the termination of *by, ham, ton,* &c). The names are written from North to South, contrary to the method observed in other maps; and there are at least twice as many names as in the others. Those of counties, or tracts of country, are generally, if not always, written within parallelograms. The roads are marked by lines; and even the miles in each stage. The rivers, like the sea, are green; and their several sources represented circular.' The Gough map measures three feet six and a half inches by one foot seven and a quarter inches.

While all these developments were taking place, similar advances were being made in the production of sea charts, which were becoming more and more important as world exploration widened. Navigation—the craft of sailing a ship between given places safely and properly—is as old as civilised man. It was well developed by the ancients; Timosthenes, a Greek sailor and scholar, wrote a work of sailing directions in ten books entitled *On Ports*, and this was copied by Eratosthenes. But the first charts—they were made by Italians and Catalans—appeared between 1200 and 1250. They were from the first more accurate than land maps, for they were always made from direct observations and by the aid of the mariner's compass or 'Stella Maris'. Sea charts are sometimes called portolans, but this is a misnomer, for the true portolan is a harbour book, or written sailing instructions, something like the itineraries issued today by the A.A. and R.A.C. for motorists. True portolans are very rare, only about twenty being known for each century of the Middle Ages. The oldest known specimen belongs to the twelfth century. This is a description of a route in the *Ecclesiastical History* of Adam of Bremen, from the mouth of the River Maas in the Netherlands to Acre in the Holy Land. In England portolans, or portolanos, became known as 'ruttiers' or 'rutters of the sea', from the French *routier*, a typical Anglo-Saxon adaptation. Some of them contained profiles of the coasts described, showing recognisable landmarks, and in time these were used also on charts.

The most characteristic feature in the appearance of a chart is the straight intersecting lines forming a kind of trellis pattern over the whole of its surface. Sometimes they are drawn in contrasting colours, sometimes in alternate continuous and dotted lines. At their intersections were, particularly in later examples, wind roses, a device invented by the Greek Timosthenes and based on the mariner's compass. These usually had their main point towards the north, and the other main compass points, or 'winds', were drawn in. These could be subdivided as many times as desired, but always in multiples of four (based on the main compass points, north, south, east and west); usually the number did not exceed thirty-two. It was from these points that the intersecting lines were drawn. They were lines of constant bearing, known as rhumb lines or loxodromes. Each represented a wind by which a ship could be sailed. They provided a mariner with a rough guide for plotting his course. If he wished to sail from port A to port B, he would lay his ruler on the chart from one to the other, and then find a loxodrome as near as possible parallel to the ruler's edge. Having found it, he traced it to its compass rose and from this he would be able to ascertain which wind to sail by.

It has always been the practice, and it is to be seen in the earliest surviving specimens, to provide charts with scales. On these, each division was subdivided into fifths, but no unit of length was indicated.

The oldest surviving chart is the *Carta Pisana* of *circa* 1300, now in the Bibliothèque Nationale at Paris. It shows the area from southern England to the Black Sea. More or less similar areas are covered by other early charts, such as those by Perrinus Vesconte (1327) and Angellino de Dalorto (about 1325).

Attempts to include Asia in world maps were being made by European cartographers by the thirteenth century. But the most important development in this extension of cartography was made during the first half of the following century by a school of cartographers, mainly Majorcans working in Catalonia, who produced a number of world maps. These cartographers became leaders in their craft partly because of their nation's maritime and commercial spirit, partly because of an influx of scholarly Jewish refugees from Berber persecution, who brought with them Arab geographical and astronomical learning, and were encouraged in map-making by the House of Aragon.

The most famous *mappa mundi* made by these cartographers is the Catalan Atlas (made before 1375), now in the Bibliothèque Nationale at Paris; it was presented by King Peter of Aragon to an emissary of the King of France. A translation of its full title runs: 'Mappamundi, that is to say, the image of the world and of the regions on the earth and of the different people which inhabit it.' Its creator is supposed to have been Abraham Cresques, a Jew living at Palma, and 'master of *mappae mundi* and of compasses' to the King of Aragon.

The Catalan Atlas is a large map. It is mounted on twelve folding screen-like boards, which when extended measure 3·9 metres in length and 69 centimetres in height. But of the twelve boards only eight are occupied by the map; the remainder contain geographical information. It is thought that Cresques was ordered to make a map showing as much as possible of both the eastern and the western extremities of the world and that this led to its elongated shape, and the exclusion of the extreme northern and southern regions. It is the earliest example of a mediaeval map showing Asia in its approximately true form, but even here it falls far short of the actuality. The huge island of Taprobana is shown, as in Ptolemy, and there are over 7,500 'spice islands' off the coast of Cathay. Spice islands there are in that part of the world—they are now known as the Moluccas—but they are certainly not so numerous as, nor in the form shown on the Catalan Atlas. Some of the details appear to have been taken from the journeys of Marco Polo, the Venetian traveller who went to China in the thirteenth century.

Other well-known Catalan world maps are the one known as the Borgia map, and one in the Biblioteca Estense at Modena, known as the Este map. The Borgia map is so called because it was bought by Cardinal Borgia for his museum at Velletri after it had been discovered in an antique shop. It is circular and engraved on two iron plates, with the engraved grooves filled with colour. The Este map, also circular, is painted on parchment.

The Catalan cartographers' approach to their subject was scientific. They made no guesses and used only information for which they had evidence. It so happened that some of the evidence was untrustworthy, but this was no fault of theirs. They were much in advance of their contemporaries.

Some important Italian maps were made at the same time as the Catalan maps. They show strong Ptolemaic influence. One such is the world map of Andreas Walsperger, a Benedictine from Salzburg, drawn at Constance in 1448. It is not a good map, but it is better than much other monastic work produced at the same time, and it contains some interesting features. The Indian Ocean is not completely landlocked, but joined by a strait with the western ocean. Infidel cities are marked by black dots, Christian cities by red ones. Taprobana is there, though it is called 'Taperbana', and reference is made to its pepper trade.

Another Italian map of about the same time is the 'Genoese' world map of 1457, now in the Biblioteca Nazionale Centrale, Florence. It is oval and is drawn to scale. It was once thought to have been made by the great Florentine scholar and cosmographer, Paolo da Pozzo Toscanelli (1397–1482), but the attribution is no longer accepted. It is based on Ptolemy, but information regarding the eastern part was taken from the then recent (1444) narrative of the Venetian traveller, Nicolo Conti, who had returned to Italy after travelling in the east for a quarter of a century.

Further developments in the portrayal of Asia were made on a *mappa mundi* made by Fra Mauro (d. 1460), a Camaldulian monk from the island of Murano (now famed for its blown glass) in the Venetian lagoon. This gave more information about the spice islands and their trade and also contained details of Java and Sumatra, and more accurate details than had appeared hitherto of China. Mauro describes Sumatra as 'Java minor'. It is, he writes, a fertile island, surrounded by eight others. In Java grow large amounts of ginger and other spices. When they are harvested they are taken to Java major, 'a very noble island' where they are divided into three parts for export. One part is carried to the north by way of the Sea of Cathay; one part is sent to Cathay; one part is sent by way of the Indian Ocean to Mecca, Jidda and Ormuz. On Java major spices are also grown, as well as much else. Gold, too, is found there. Yet, he writes, despite its riches and pleasant scenery, the people who inhabit it are idolatrous and evil.

Japan, which is called Zimpangu, appears here for the first time on a European map, but it is distorted into a small island north of Java. Africa, too, is shown in greater detail than on previous maps; Mauro probably obtained his information for this from members of the Coptic church from Ethiopia visiting Venice. Jerusalem is abandoned as the world's centre. The map is oriented with the south at the top.

The Mauro map, which was commissioned by the King of Portugal and finished on 24 April 1459, was not solely by Mauro, for he had the assistance of Andrea Bianco, a chart maker, and of several draughtsmen and illuminators. The monastery accounts contain details of payments for materials and to craftsmen in connexion with Mauro's work. The original map has disappeared, but there is a contemporary copy of it, probably by Bianco, in the Biblioteca Marciana at Venice. It is circular and measures about six feet four inches in diameter. A contemporary medal struck in Mauro's honour describes him as 'geographus incomparabilis'. As in the Catalan Atlas, much of the detail in Mauro's map appears to have come from Marco Polo.

There was a great revival of interest in Ptolemy's *Geography* during the fifteenth century. This was partly because in 1400 a Florentine, Palla Strozzi, brought back from Constantinople a copy of the 'A' recension of the work. A translation by Jacopo Angelus and Manuel Chrysolorus was completed by 1406, and by four years later the Florentines, Francesco di Lappacino and Domenico de Boninsegni, had re-drawn the maps with Latin legends. Many manuscript copies of Ptolemy followed this, some of them attempting to indicate relief by various methods. The most successful of these was made by Henricus Martellus, now in the Biblioteca Nazionale in Florence; it depicts the relief of the Alps by means of elaborate shading and colour. Another Florentine, Francesco Berlinghieri (1440–1550), translated the *Geography* into Italian verse about 1480. It was in manuscript and was not printed until 1482, by Maestro Nicolo Todesco. Berlinghieri added four new maps to the work, of Palestine, Gaul, Italy and Spain, and these were the first maps of the modern world to be published in an edition of Ptolemy. He also revised several of the old maps.

Printing made possible the distribution of maps on a much wider scale than hitherto, and many people now saw maps who had not seen them before. Printing, with its kindred craft of engraving, also enabled them to be repeated with precision. Before this, hand-copying had subjected cartography to a multitude of variations. Maps were liable to corruption every time they were copied by hand; in printing they were liable to corruption at only the first stage. Yet this, too, had its drawbacks, for an error once made was usually perpetuated. Many of the earlier *mappae mundi* in old manuscript works were reproduced in print more or less as they were, as for example the T-O map from a work by St. Isidore of Seville (1482; page 17).

The earlier editions of Ptolemy were printed in limited numbers.   The first printed at Bologna in 1477[1] by Dominico de' Lapi was limited to five hundred copies, which sold slowly. It contained twenty-six maps by the miniaturist Taddeo Crivelli (d. *circa* 1484), painter of the miniatures in the great Bibbia Estense of Duke Borso of Ferrara, who was by 1477 employed at the court of Giovanni Bentivoglio at Bologna. The maps are drawn, as in the manuscript editions, on a conical projection, and show climates and degrees of longitude and latitude in the margin.   It is not a good edition, despite the fact that it was the first book to contain printed maps, for the maps themselves are inaccurate and incomplete and the text full of misprints.   It is said to have been rushed through the press to forestall a Rome edition of 1478.   It is also said that a workman was enticed by de' Lapi from the Rome printing house and persuaded to reveal methods used there.

There were five further editions of Ptolemy before 1500.   They were those of 1478 (Rome: printed by Conrad Sweynheym and Arnold Buckinck); 1482 (Florence: printed by Nicolo Todesco; this is the Berlinghieri edition already noted); 1482 (Ulm: printed by Lienhart Holle); 1486 (Ulm: printed by Johann Reger); and 1490 (Rome: printed by Petrus de Turre).   All, with the exception of the Ulm editions, have copperplate maps; the Ulm maps are woodcuts.   Of these, the 1478 edition has its maps engraved by Sweynheym on a rectangular projection, with 'Trajan column' lettering.   Degrees of longitude and latitude are shown in the margins, the duration of the longest day is indicated, and mountains are shown in elevation, like molehills.   The first Ulm edition (1482) had its maps engraved on a trapezoidal projection; there were thirty-two of them, five of them of the modern world.   It was the first edition of Ptolemy to contain a map of Greenland.   It also contained the first world map to show the great discoveries of the time, and was the first map to be signed by the engraver, who was Johanne Schnitzer de Arnescheim.   The next Ulm edition (1486) contained additions to the map of Germany.

Editions of Ptolemy continued to be printed after 1500.   Of these, an edition printed by Bernard Sylvanus at Venice in 1511 contained twenty-seven woodcut maps in black and red, and a heart-shaped world map.   Another edition, one of the finest of all, was printed at Strasbourg in 1513.   The 1511 and 1513 editions, with some names printed in red, are the only recorded instances of *printed* colour being used on maps before the nineteenth century.

---

[1] The date is misprinted on the colophon as 1462.   An edition without maps had been printed at Vicenza in 1475.

The 1513 edition contained forty-seven woodcut maps. It was begun by Martin Waldseemüller, but finally published by Jacobus Ezler and Georgius Ubelin. After this edition the blocks were used again in editions at Strasbourg in 1520, 1522 and 1525, at Lyons in 1535, and in Dauphiné in 1541.

The maps of the Ptolemy editions were not the only ones to be printed in the fifteenth century. Most fifteenth-century printed maps were published in books, for there was virtually no demand for separate engraved maps, although a few were produced, particularly later in the century. Other books of the period containing maps include Hartmann Schedel's *Liber cronicarum* (*Nuremberg Chronicle*) of 1493, which contains maps of Germany and of the world. There are also Bartolomeo dalli Sonetti's *Isolario* (Venice 1485), the *Rudimentum novitorium* (Lübeck 1475), which contained maps of Palestine and the world, and the *Peregrinatio in Terram Sanctam* by Bernhard von Breydenbach (Mainz 1486). Some of these maps were of great importance because they were used as prototypes by later cartographers. This is true of the maps of Francesco Roselli (d. 1513) of Florence, the earliest known commercial mapseller. He produced and sold maps from 1482 until he died. His son carried on his business, and when he died in 1527 an inventory of his shop was taken, which listed all kinds of things, from charts to views, from gores[1] for globes to *mappae mundi*. All are rarities today.

BIBLIOGRAPHY

BAGROW, LEO: *History of Cartography*. Revised and enlarged by R. A. Skelton. 1964.
—— 'The origin of Ptolemy's "Geographia"'. *Geografiska Annale*, Vol. XXVII, pp. 318–87. 1945.
BEAZLEY, C. RAYMOND: *Dawn of Modern Geography*. 3 vols. 1897–1906.
BROWN, LLOYD A.: *The Story of Maps*. 1951.
BUNBURY, E. H.: *History of Ancient Geography*. 2 vols. 1879.
DAVENPORT, WILLIAM: 'Marshall Islands Navigational Charts'. *Imago Mundi*, Vol. XV, pp. 19–26. 1960.
GUZMÁN, E.: 'The Art of Map-making among the ancient Mexicans'. *Imago Mundi*, Vol. III, pp. 1–6. 1939.
JACOBS, J.: *The Story of Geographical Discovery*. 1899.

[1] Gores are the sections of a map applied to a sphere in making a globe.

KIMBLE, G. H.: *The Catalan world map of the R. Biblioteca Estense at Modena.*   1934.
—— *Geography in the Middle Ages.*   1938.
MYRES, J. L.: 'An attempt to reconstruct the maps used by Herodotus'.   *Geographical Journal*, Vol. VIII, pp. 605–31.   1896.
NORDENSKIÖLD, A. E.: *Facsimile Atlas to the early History of Cartography.*   1889.
STEVENS, HENRY: *Ptolemy's Geography.*   1908.
TAYLOR, E. G. R.: *The Haven-Finding Art.*   1956.
THOMSON, J. O.: *History of Ancient Geography.*   1948.
UNGER, E.: 'From the cosmos picture to the world map'.   *Imago Mundi*, Vol. II, pp. 1–7.   1937.
WARMINGTON, E. H.: *Greek Geography.*   1934.
WATERS, D. W.: *The Rutters of the Sea.   The Sailing Directions of Pierre Garcie.   A study of the first English and French sailing directions . . .*   1967.

CHAPTER TWO

The Dawn of Modern Cartography

THE great fifteenth- and sixteenth-century voyages of discovery speeded the development of cartography and made it more international in outlook. Until this time most original research had been done by Spanish and Italian geographers, but now cartographers of other nationalities began to have an ever-increasing influence, in particular the Dutch, French, Portuguese and English. These peoples, and the Spanish and Italians, had sent out voyages of discovery to most parts of the world, and a vast amount of information had been brought back on which cartographers could base improvements and extensions of their maps. Diaz, Magellan, Columbus, Vasco da Gama, Cabral and others opened up new areas undreamed of by many of their predecessors and contemporaries.

In the developments which made these discoveries possible—better ships, greater meteorological knowledge, more refined methods of navigation—the Portuguese were pre-eminent. This helped to improve both the content of their maps and the technique of their cartography. The Portuguese contribution to the development of cartography applied in particular to Africa, for Portuguese navigators did much towards the charting of the African coastline. At the zenith of their discoveries occurred Bartholomew Diaz's rounding of the Cape of Good Hope, and Vasco da Gama's opening of the sea route to India in 1497–9.

But only one or two Portuguese charts of the fifteenth century have survived, and for much of this information we have to look to contemporary Italian, and especially Venetian, maps which incorporate details taken from Portuguese sources. One interesting Venetian chart of the time is in the British Museum (Egerton MS. 73). This records exploration by the Portuguese navigator Diogo Cão in the years 1482–4, as far as the Congo (Plate 3). It is entitled 'Ginea Portugalexe'. The book in which it is bound contains also maps of further discoveries by Cão and of Diaz's voyage round the Cape.

Portuguese navigation is also involved with the story of Christopher Columbus, for it is said that he thought of the idea of a westerly passage to Cathay and the Indies while serving on a Portuguese ship. One map has survived which was drawn by Columbus on his voyages. It is a rough sketch of the coast at the north-west corner of Hispaniola (now Haiti and the Dominican Republic), made in December 1493, and now in the collection of the Duke of Alba. A handful of other maps having close connexions with Columbus survives. One of them is in the Museo Naval at Madrid. It is a world map, dated 1500, by Juan de La Cosa, who sailed with Columbus on his 1493–4 voyage. It represents the first three voyages of Columbus, and that of John Cabot in 1497. There are, too, some sketch maps made by the Italian geographer Alessandro Zorzi, illustrating Columbus's own ideas of the nature of his discoveries, showing them as part of Asia; for Columbus never believed he had discovered a new continent. It was once thought that these maps, which are now in the Biblioteca Nazionale at Florence, were drawn by Columbus's brother Bartolomeo, but they have now been ascribed to Zorzi.

It was some time before cartographers accepted that a new continent had been discovered, and the first one to show it as a separate entity was Martin Waldseemüller, in his world map of 1507. Waldseemüller was born at Radolfzell on Lake Constance (Boden See) about 1470. He was educated at Freiburg. He made his home in Lorraine at St. Dié, then a centre of learning, and there worked on an edition of Ptolemy. His world map—a woodcut in twelve sheets published in an edition of one thousand—was published in 1507. Its title acknowledges the cartographer's debt to the ancient learning of Ptolemy and to the recent discoveries by Amerigo Vespucci, the Italian navigator who had sailed with Columbus in 1499: *Universalis Cosmographia secundum Ptolomaei traditionem et Americi Vespucci aliorumque lustrationes*. Indeed it was Waldseemüller who first suggested, in his *Cosmographio Introductio* of 1507, that the new continent should be named America after Amerigo Vespucci, who had written a widely-circulated account of the natives of Venezuela.

Only one impression of Waldseemüller's world map has survived; it is in the Schloss Wolfegg, in Württemberg. He also issued a *Carta Marina* in twelve sheets; the only impression of this is also in the Schloss Wolfegg, bound with the world map. There are several reasons for these maps having survived in such small numbers or in single impressions. The sheets were usually stuck together on canvas or cloth, to be used as wall hangings. In this form they were too large for glazing, and whether they were rolled up and stored in cupboards or simply left on the wall, they cracked, became torn, or rotted away.   Nearly all those that have survived are in book form.   In fact one of the reasons why atlases came into being was because of the ephemeral quality of wall maps; the single sheets bound as books were much more durable.   At least twelve sixteenth- and early seventeenth-century maps have survived in only single impressions.   Some which have not survived are known only by repute.

The *Carta Marina*, or to give it its full title, *Carta Marina Navigatoria Portugallen Navigationes* was derived from a Portuguese source and based on Portuguese voyages. A reduced copy of the chart was published by Laurent Fries in three separate editions, in 1522, 1527 and 1530.   Only one copy of the 1530 edition has survived.

Many of the discoveries made at this time were slow in being recorded on maps. Cartographers at first tried to fit the new discoveries into existing outlines, but in due course more scientific methods prevailed.   The first map so conceived had ante-dated Waldseemüller's world map by a year.   It was engraved on copper in 1506 by Francesco Roselli after the design of Giovanni Matteo Contarini.   It is a map of the northern hemisphere, with the North Pole at its centre, and drawn on a conical projection (Plate 5).   It is possible that another map was made of the southern hemisphere.   The main source for this map of the northern hemisphere was the maps of Ptolemy, to which were added corrections in the light of the new Portuguese discoveries.   It was the first stage in the development of modern cartography—a stage that was completed by the publication of Waldseemüller's world map and *Carta Marina*.

There was a large increase in demand for maps and charts in the sixteenth century, and publishers were not slow to meet this demand by publishing maps which embraced all the latest information.   Two of the greatest of such publishers were the Flemings, Mercator and Ortelius.

Gerardus Mercator is the Latinised form of the Flemish name Gerhard Kramer.   He was born in 1512 in Rupelmonde in East Flanders, and was of German descent.   He studied at the University of Louvain under Gemma Frisius, the astronomer of the Emperor, and inventor of an astronomical ring dial by which a mariner could tell the time by the sun in any latitude.   Mercator engraved a set of globe gores for Frisius in 1536.   He was patronised by the Emperor Charles V (who, in common with other contemporary princes, had a room hung entirely with maps), and then in 1559 became cosmographer to the Duke of Jülich and Cleves.   Charles V's patronage brought Mercator in contact with the great Portuguese and Spanish cartographers and navigators and consequently with the latest geographical developments.   He was something of a polymath—clever in mind and hand: *ingenio dexter, dexter et ipse manu* was how his contemporaries described him—for in addition to much else, he was a land surveyor, a scientific instrument maker and an engraver.

Mercator's map projection is still used today. It was first used on his twenty-four sheet world map published in 1569, of which only four impressions still exist. Its main feature is that the parallels are placed at increasing intervals from the equator, in proportion to the increasing distances between the meridians; the parallels are thus known as 'waxing latitudes'. The projection was particularly welcomed by mariners because it enabled them to take a straight line as constant bearing. Other contemporary maps and charts made no allowance for the convergence of the meridians.

But quite apart from his projection, Mercator introduced many refinements and improvements into cartography. Among these were the reduction in length of the Mediterranean to nearer its true size, a more correct position for the Canary Islands, and a more correct representation of the European land area between the Baltic and Black Seas. All of these were included in Mercator's map of Europe published in 1554 at Duisburg, during Mercator's term there as a university lecturer. More improvements were included in a second edition of the map published in 1572. Other errors had to wait longer for correction: he distorted the shape of America, for example, and showed Tierra del Fuego joined to the Antarctic continent. But Mercator was a conscientious cartographer and made improvements wherever he could.

Mercator's world map was part of his plan to publish a collection of maps of various parts of the world. He published a map of the British Isles in 1564, but the collection proper had its first part published in 1565 at Duisburg, when he was fifty-three. There were fifty-one maps, in three parts, each part with a separate title page: Belgium, France and Germany. A second group followed in 1590, consisting of twenty-two maps of Greece, Italy and Slavonia. Finally a third part followed a year after Mercator's death, in 1595; it consisted of thirty-six maps. The work was then published as a whole, with a general title page, on which the world 'atlas' was used for the first time for a collection of maps: *Atlas sive cosmographicae meditationes de fabrica mundi et fabricati figura*. There were some forty-seven editions of the *Atlas*, with the text in various languages, between 1595 and 1642.

Mercator's maps, which were engraved on copper and issued plain or handcoloured, are artistic and notable for their fine italic lettering. Lettering was another of his achievements; he published a treatise on the subject in 1540 at Antwerp. Hills and mountains are shown in elevation and towns are symbolised by groups of buildings.

In places Mercator depicts exotic animals such as elephants and camels, all correctly drawn. Correct, too, are his representations of ships sailing on the seas (Plate 4). But, like many other cartographers of his time, he could not resist in places the insertion of sea monsters. Perhaps it was the sight of such cartographic monsters that moved Edmund Spenser to write in *The Faerie Queene*:

> Most ugly shapes and horrible aspects,
> Such as Dame Nature selfe mote feare to see,
> Or shame that ever should so fowle defects
> From her most cunning hand escaped bee;
> All dreadfull pourtraicts of deformitee:
> Spring-headed Hydres; and sea-shouldring Whales;
> Great whirlpooles which all fishes make to flee;
> Bright Scolopendraes arm'd with silver scales;
> Mighty Monoceroses with immeasured tayles.
>
> The dreadful Fish that hath deserv'd the name
> Of Death and like him lookes in dreadfull hew;
> The griesly Wasserman, that makes his game
> The flying ships with swiftnes to pursue:
> The horrible Sea-satyre, that doth shew
> His fearfull face in time of greatest storme;
> Huge Ziffius, whom Mariners eschew
> No lesse than rockes, (as travellers informe)
> And greedy Rosmarines with visages deforme.

Mercator had several sons, of whom Arnold (1537–87) followed in his father's footsteps as a cartographer. He made maps and town plans. Another cartographer son was Rumold, who compiled the world map for the first collected edition of the *Atlas*. Arnold's sons, Gerhard, John and Michael, also became cartographers; John holding the post of cartographer to the Duke of Cleves. They all contributed to various editions of the *Atlas*.

We must now consider Abraham Ortelius, this being the Latinised form of his Flemish surname, Wortels. He was born at Antwerp in 1527, the son of Léonard Wortels, who was probably an antique dealer. Certainly Abraham had a great interest in such things, and himself formed a valuable collection of works of art. Léonard died young, while Abraham was still a child, and the boy was brought up by an uncle. Little is known of his youth, apart from the fact that he was a serious student, interested particularly in history and geography. He had two sisters to support and in order to do this adopted, with his sisters as assistants, the trade of a mapseller and map colourist. He was so registered by the Guild of St. Luke in Antwerp in 1547. He made many influential friends in the business and artistic worlds. He travelled a great deal in Italy, France and Germany, and visited England, where he helped to persuade William Camden to produce his *Britannia*. He knew the great English geographer, Richard Hakluyt. In 1560, Ortelius travelled through France with Gerhard Mercator. It is thought that it was on this trip that the foundations were laid of his own original work in cartography.

The maps produced by Ortelius before 1570 are largely derivative. In his world map of 1564, for instance, he used a projection based on one already used by the German cartographer, Peter Apian in 1530 (engraved in 1540), and his map of Asia (1567) was derived largely from the work of the Venetian cartographer, Jacopo Gastaldi. But it was not so much for his separate maps that Ortelius was noted, as for his great and successful work, *Theatrum Orbis Terrarum* (1570), which was published by Aegidius Coppen Diesth at Ortelius's expense (Plate 6). It was the first modern atlas, though it was not known by that name, and was an unprecedented achievement. Ortelius received in 1573 from King Philip II the title of Geographer to the King. Originally the *Theatrum* contained seventy engraved maps, of uniform size and style, on fifty-three sheets. There were a world map, four maps of continents, and sixty-five regional maps; three of Africa, six of Asia, and the remainder of Europe. It was issued in colours or plain. From 1579 Ortelius added to this work a collection of ancient maps entitled *Paregon Theatri*.

The *Theatrum* is a work based on wide authority, and Ortelius acknowledges this. In the first edition he listed eighty-seven cartographers, and by 1603 there were a hundred and eighty-three. It was a popular and successful work, as may be deduced from the fact that between 1570 and 1612 over forty separate editions were published in various languages. Apart from his maps, Ortelius published works on coins, place-names, topography and other subjects.

Ortelius was not a geographer in the same sense as Mercator. He was a scholarly craftsman and business man who produced attractive maps based on information derived from Mercator and others. His maps met a current demand for such work and had the success they deserved. The many editions of the *Theatrum* made Ortelius's fortune, and he was able in 1581 to buy a new house, De Vlasbloem (the Flax-flower) in the rue de l'Hôpital, where he settled with his sister Anne and their aged mother. Here he enlarged the *Theatrum* and enjoyed his splendid collection of *objets d'art*. Later he moved again into a house in the Rue de Couvent, taking over the next-door house as well, and knocking them into one.   Here he established the 'Museum Ortelianum' which became well known.   He died on 4 July 1598.

Among the engravers who worked for Ortelius on the *Theatrum Orbis Terrarum* was Frans Hogenberg (*circa* 1540–*circa* 1590), a member of an old family of Mechlin artists. He lived for a long time at Antwerp, but was driven thence by Spanish oppression, after which he settled in Cologne.   Here he met Georg Braun or Bruin (1541–1622) and another refugee from Mechlin, Simon van den Neuvel.   They published a city-atlas with a text by Ortelius and Cornelis Caymox, the *Civitates Orbis Terrarum*, issued in 1572 at Antwerp.   It is sometimes found bound with the Ortelius *Theatrum*. There were many editions, in Latin, German and French.

One of the more important artists who took part in the *Civitates* was Georg Hoef-nagel (1542–1600) a native of Antwerp.   Most of the 'maps' in the work are bird's-eye views, but Hoefnagel's are taken in frontal elevation, and are greatly detailed.   The surrounding landscape is also shown.

France, too, was influenced by the new world discoveries.   Portuguese navigators and chartmakers served in French ships and their influence was particularly strong in Normandy.   Some noted French cartographers worked at Dieppe, among them Jean Rose or Rotz, who made an atlas of the coasts of Europe, America and Africa for the English king, Henry VIII; Nicolas Desliens, who made two world maps in 1541 and 1566; Pierre Desceliers, who made a world map in 1546; and Nicolas de Nicolay of Dauphiné, who visited England in 1546–7, and made charts of the British coast and a map of Scotland.   Oronce Finé, or to give him his Latinised name, Orontius Finaeus Delphinas (1494–1555) published in 1519 a world map in a single heart-shaped pro-jection.   Later, in 1531, he made one in a double heart-shape for *Novus Orbis regionum* by Simon Grynaeus and John Huttich.[1]

Italy, where the sixteenth-century map trade was in its most developed state, had some fine cartographers, including the Freducci family, who worked at Ancona from 1497 to 1566; the Maggiolo family, who worked at Genoa from 1504 to 1588; and the Olives family, who worked first in Majorca and then at Marseilles from 1532 to 1659. There was, too, Battista Agnese of Venice, (*fl.* 1536–64) who produced over sixty atlases which included single-country maps and charts.   An inventory of one map dealer, Francesco Rosselli of Florence, showed that he carried thousands of maps in stock.

One of the greatest and most influential of sixteenth-century Italian cartographers was Giacomo Gastaldi (*circa* 1500–*circa* 1565), particularly famous for his oval world

[1] *See also* page 111.

10. Title page from Waghenaer's *Spieghel der Zeevaerdt*, 1585. *British Museum.*

11. Map of South America, engraved by A. F. van Langren, 1595.

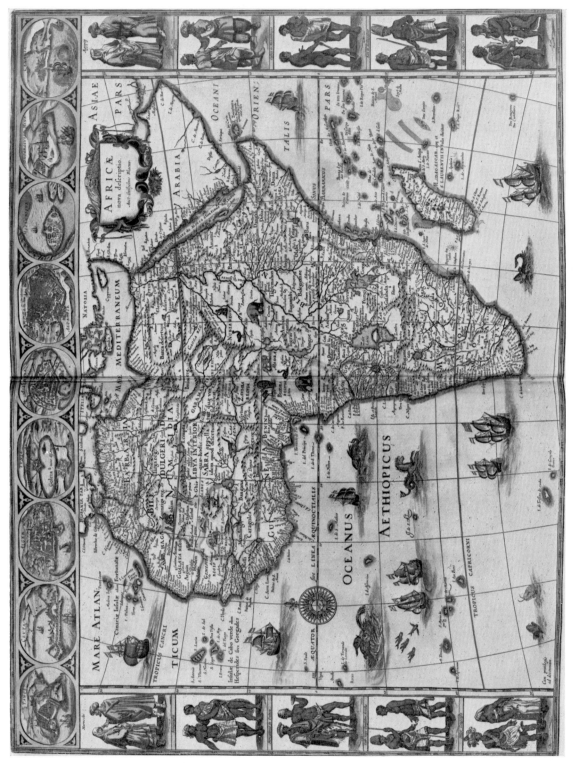

12. Map of Africa, by G. Blaeu, from *Grooten Atlas*, 1648–65.

British Museum.

13. Hondius: Sir Francis Drake's Map of The World, *circa* 1590. *British Museum*

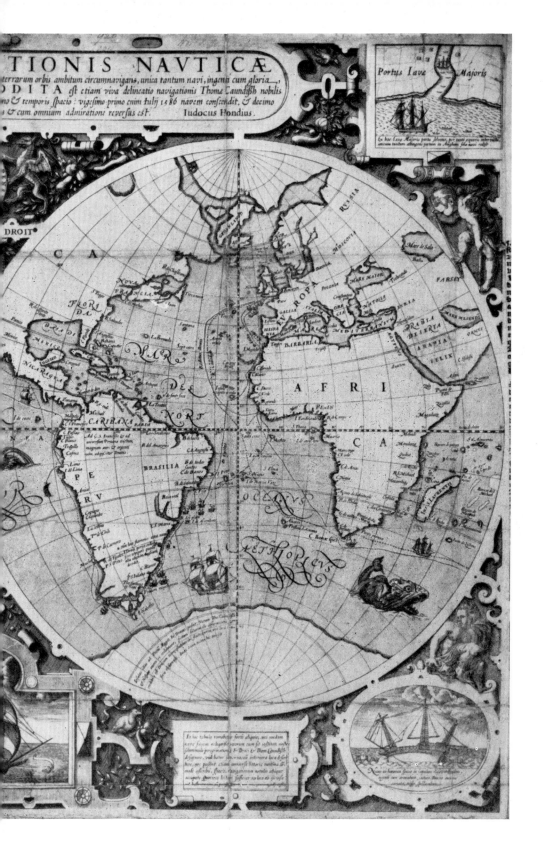

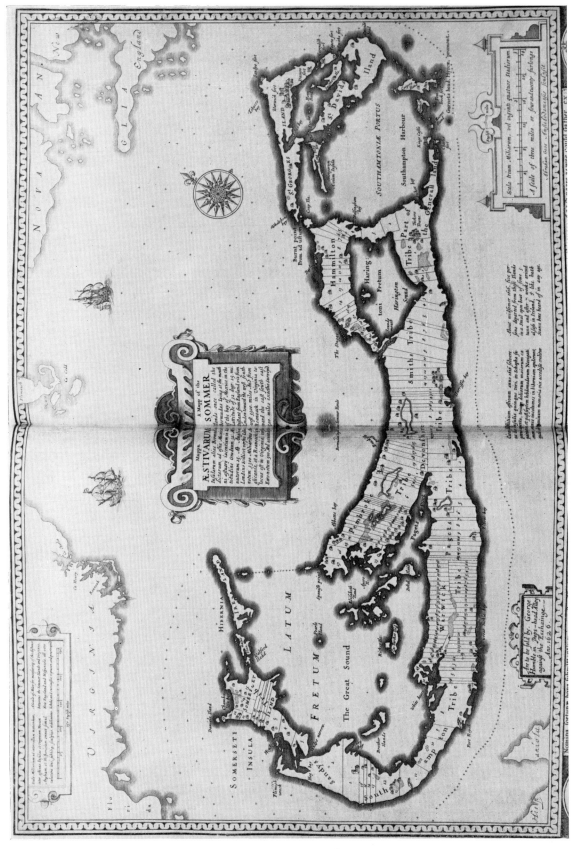

14. Map of Bermuda by G. Blaeu, from *Atlas Maior*, 1662.

British Museum.

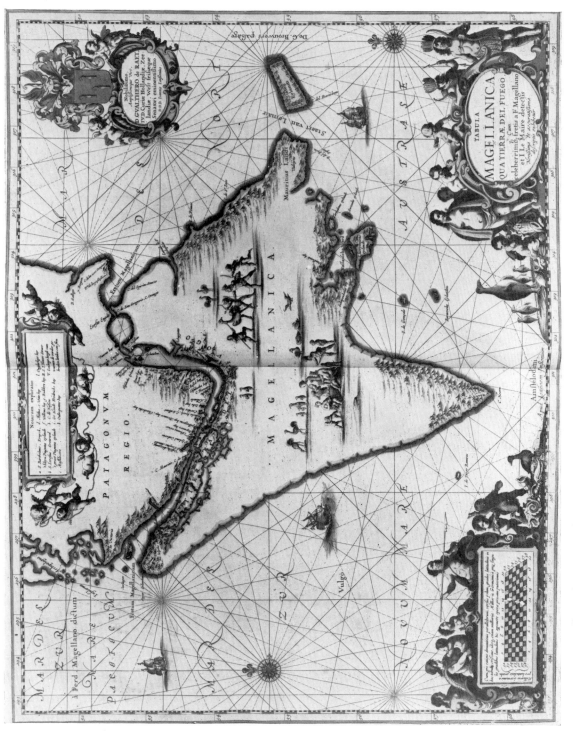

15. *Magellanica* by Jansson from *Novus Atlas*, 1658.

British Museum.

16. Title page of Jansson's *Novus Atlas*, Vol. IV, 1659.   *British Museum*.

map of 1546, which was extensively copied.   Altogether he made nearly a hundred maps between 1544 and his death (Plate 7).   Gastaldi was also an engineer in the service of the Venetian Republic.   Among map publishers as distinct from true cartographers in Italy, was Antonio Lafreri (1512–77), a Frenchman whose real name was Antoine Lafréry.   In addition to publishing maps he carried a large stock of work published by others.   He made collections of individual maps to customers' requirements and had them bound into books, each with an impression of his special title page. As they were made up individually, these collections are rarely alike; certainly their contents are not standardised.   To make up such atlases was a common practice in those days, especially in Rome and Venice.

In Germany at this period there were three main centres of cartography: Nuremburg, Vienna and the Rhineland.   Each centre produced outstanding cartographers.   One of the most famous of all was Peter Apian or Bienewitz (Plate 9).   He had a printing press at Landshut, and in addition to his cartographic and printing activities was a professor of mathematics.   His book *Cosmographicus Liber* went through sixty editions between 1524 and 1609, the later ones being edited by Gemma Frisius, his pupil. Other notable German cartographers of the period included Caspar Vopel of Cologne, who in 1545 produced a world map in twelve sheets; only one incomplete copy of the first edition survives.   He produced another in ten and a half sheets in 1555, and no copy of the first edition of that has survived.   Later editions of both are known, but are of great rarity.   Johann Hunter of Basle published in 1530 his *Rudimentorum Cosmographicae libri duo*, a work similar to Apian's *Cosmographicus Liber*.   This contained two maps, but a later edition, in 1542, had sixteen.   Numerous editions were subsequently published by the Basle publisher, Henricpetri.   Sebastian Münster, also of Basle, published in 1540 *Geographia Universalis*, with subsequent editions in 1542, 1545, 1551 and 1552; it probably influenced Ortelius.   This was followed in 1544 by *Cosmographia*, which contained twenty-six maps.   Like the earlier work it was reprinted several times (cf. Plate 51).

BIBLIOGRAPHY

BAGROW, LEO: 'Essay of a Catalogue of Map-incunabula'.   *Imago Mundi*, Vol. VII.   1950.
—— *History of Cartography*.   1964.

c

CONTARINI, G. M.: *A map of the world designed by G. M. Contarini, engraved by F. Roselli, 1506.* 1924 (British Museum).

HARRISSE, H.: *The discovery of North America.* 1892.

HEAWOOD, E.: 'The world map before and after Magellan's voyage'. *Geographical Journal,* Vol. LVII, p. 431. 1921.

—— 'A hitherto unknown world map of A.D. 1506'. *Geographical Journal,* Vol. LXII, p. 279. 1923.

KEUNING, JOHANNES: 'The "Civitates" of Braun and Hogenberg'. *Imago Mundi,* Vol. XVII, pp. 41–4. 1963.

KISH, GEORGE: 'The cosmographic heart: cordiform maps of the 16th century'. *Imago Mundi,* Vol. XIX, pp. 13–21. 1965.

KOEMAN, C.: *The History of Abraham Ortelius and his Theatrum Orbis Terrarum.* 1964.

OSLEY, A. S.: *Mercator.* 1969.

POPHAM, A. E.: 'Georg Hoefnagel and the Civitates Orbis Terrarum'. *Maso Finiguerra,* anno I (1936), pp. 183–201.

SKELTON, R. A.: 'The cartography of Columbus' first voyage'. *Vigneras, L. A., Journal of Christopher Columbus.* 1960.

TOOLEY, R. V.: 'Maps in Italian atlases of the sixteenth century'. *Imago Mundi,* Vol. III. 1939.

UHDEN, R.: 'An unpublished Portuguese chart of the New World, 1519'. *Geographical Journal,* Vol. XCI, pp. 44–56. 1938.

—— 'The oldest Portuguese original chart of the Indian Ocean, 1509'. *Imago Mundi,* Vol. III, p. 7. 1939.

CHAPTER THREE

The Low Countries

T HERE have been few periods, and few places, in which the craft of mapmaking reached such heights as it did in the late sixteenth, and in the seventeenth centuries in the Low Countries. The maps produced by early Netherlandish cartographers combined accuracy with sumptuousness of presentation. These two properties have not always been brought together.

I have already written in Chapter Two of the great influence of Mercator and Ortelius. But there were other contemporary cartographers in the Low Countries who were helping to build up the great reputation to be enjoyed by Dutch and Belgian mapmakers in the late sixteenth and seventeenth centuries. Such were Jacob van Deventer (*fl.* 1540–5) who worked with the great Dutch publisher Christopher Plantin, and Cornelis Anthoniszoon or Anthonisz (1499–1556?), who was well known for his charts of the Zuyder Zee and of the Baltic and Scandinavian seas. Cornelis Anthoniszoon was a versatile man, for besides being a cartographer he was also a navigator, painter, draughtsman and engraver. In 1531–2 he made a manuscript plan of Amsterdam, the earliest one on record, which was engraved on wood and published in 1544. A year before this Anthoniszoon had published his 'Caerte van Oostlant', the work for which he is most famous. No copy of the first edition is known. It was really a sea chart (*paskaart*), for in an accompanying description sailing instructions are given and dangerous areas noted.

There were also Adrian Gerritsz (*circa* 1525–79), William Barentszoon (1550–97) and Govert Willemsz (*fl.* 1550), all noted for their sea charts. These, and others, helped succeeding generations of Dutch and Flemish cartographers to take the lead from the Italians.

Of Dutch chartmakers active in the early part of the great period, the most famous is Lucas Janszoon Waghenaer (1533 or 1534–1606), creator of the great sea atlas *Spieghel* der Zeevaerdt (Plate 10). He was born at the fishing port of Enkhuizen of West-Frisian parents.   He was a pilot and his work brought him into contact with Italian, Portuguese and Spanish navigators.   These men showed him their charts, which were to have a profound influence on his own work.   Waghenaer's first cartographic work was a plan of his native town, executed in 1577.   It was engraved by Harmen Janszoon Muller of Amsterdam.   Two years later his charts were engraved for the first time, and at about the same time (1579) he gave up his work as a pilot to take up a fiscal post in Enkhuizen.   Waghenaer was dismissed from this post in 1582 because, it is thought, he had accepted bribes.   Thereafter, with a large family to support (he had married about 1560 and had eight children), he was frequently short of money and had to take on part-time jobs and raise loans to make ends meet.   Moreover, he was himself financing the engraving of his charts for *Spieghel der Zeevaerdt*, which was an expensive undertaking.   The first part of the work was printed by Plantin at Leyden in 1583.   It was dedicated to Prince William of Orange; the original copy presented to him is preserved in the University Library at Utrecht.   The work was successful and its second part appeared in 1585.   The local government made Waghenaer a grant at this time 'from income from taxes on contraband and spoils', thus officially restoring his character.

In the same year Waghenaer's friend, François Maelson, was sent as Ambassador to England, and took a copy of the *Spieghel* to show to the English Privy Council.   The Latin edition of the work, which appeared in the following year, was dedicated to Queen Elizabeth.   In time this work gave a new word to English mariners, 'waggoner' (from Waghenaer), for a sea atlas.   The work was first published in English, and completely re-engraved in 1588 by A. Ashley.   It was then called *The Mariner's Mirrour*.   Features incorporated in the *Mirrour*, still used on Admiralty charts today, are silhouettes of coastlines as they appear from six to nine miles out at sea (Plate 8).

In time Waghenaer became a famous and wealthy man, and later in his life he took an active part in preparing charts from Portuguese expeditions led by the Dutchman Jan Huygen van Linschóten and Dirck Gerritsz Pomp ('Dirck China').   These were to appear in a work by H. van Linschóten entitled *Itinerario* (1596), in Waghenaer's own *Thresoor der Zeevaerdt* (1592), and in the pocket-sized 'rutter of the sea' *Enchuyser Zeecaertboeck* (1596–8) by Linschóten and Waghenaer.   Later Waghenaer again fell on lean times, for in 1601 the government granted him a pension of a hundred florins a year, and when he died in 1606 his widow was in straitened circumstances.

The first edition of the *Spieghel* was published in 1585.   It was reprinted in 1591,

1594, 1605 and 1613.   It contained a general chart of western Europe with rhumb lines, and twenty-two more detailed charts of European coastal waters.   The second part contained twenty-one charts giving soundings, shipping channels, sandbanks, buoys and beacons, though the coast is not always drawn to scale or accurately.   For the first time coastal profiles were shown on the chart itself, integral with the coastline; earlier, in portolans, they had been included in the text.   They show salient features such as large buildings, church towers, groups of buildings, hills and cliffs.   Decorations are splendid and include many kinds of ships, titanic monsters, colourful compass roses and flamboyant cartouches for titles and scales.   On these latter may be seen a rich conglomeration of putti, parrots, birds of paradise, fruit, shells, masks, jewels, strapwork, busts and cloths of estate.

The title page, in common with those of many other atlases, is magnificent.   The central motif is a Dutch cupboard with architectural details, surmounted by a convex mirror, around which are grouped half a dozen mariners.   The doors of the cupboard are replaced by a tablet for the book's title.   On each side of the cupboard is a pilot taking soundings with a lead and line.   Below is the sea on which sails a ship flanked by monsters.   Around the whole design are representations of celestial and terrestrial globes, dividers, hour glasses and other navigational instruments and aids (Plate 10).

Another Dutch cartographer of roughly the same period as Waghenaer was Gerard de Jode (1509–91), noted for his *Speculum Orbis Terrarum* (1578), an atlas in two parts containing sixty-five maps, thirty-eight of them forming the second part and all devoted to provinces of the German empire.   The maps were engraved by the brothers Jan and Lucas van Deuticum.   A revised edition of the *Speculum* was published by Gerard's son Corneliss in 1593, in which eighteen more maps were added, bringing the total to eighty-three.   Corneliss de Jode was a rival of Ortelius, who was eighteen years younger than himself, but found it impossible to make much headway against the young man, because he was so well established (Plate 7).

Jodocus (1563–1612) and Henricus Hondius (1597–1657) formed a father and son partnership.   Their real name was de Hondt—Hondius is the Latinised form.   Jodocus for a time settled in London where he worked as an engraver and typesetter, among other things engraving gores for the first English globes.   Here he married a compatriot, Collette van den Keere, whose brother assisted him in engraving maps.   In 1593–4 Hondius returned to Amsterdam and took over the business and stock of Mercator; he issued a new and enlarged version of Mercator's *Atlas* in 1606.   Beginning in 1605 he engraved a number of maps for the famous folio atlas produced by the

Englishman John Speed.   By 1609 he was the leading map publisher in Amsterdam (Plate 13).   His trade sign was *de wackere hond* (the watching dog).   When Jodocus died his sons Jodocus II (1595–1629) and Henricus took over the business, and in 1635 joined forces with Jan Jansson (1596–1664) to publish an improved edition of the Mercator/Hondius atlas.   Jansson married the sister of Henricus, and when the latter died, succeeded to the business.

Little is known of the work of Jodocus Hondius II, but we do know that he engraved maps, signing them as Jodocus Hondius Junior.   He later dropped the word Junior from his name, thus causing some difficulty in deciding whether to ascribe work to father or son.   However, we know that he made thirty-four maps that are found in Blaeu's rare *Appendix* of 1630 (*see* page 40), only one copy of which is known.   It is in the British Museum.   In 1613 he engraved a fine map of the Baltic area, *Nativus Sueciae adiacentiumque Regnorum Typus*, and published a pair of terrestrial and celestial globes, with a guide for their use published in separate editions in Dutch and Latin.   In 1618 he published a new edition of a world map in two hemispheres, originally published by his father.   Only one copy of this is known—it is in the Schloss Wolfegg. No copy is known of the father's edition of 1608.   Among the other publications of the younger Jodocus Hondius are the *Tabulae contractae* of Petrus Bertius (1616; 216 small maps); The *Theatrum Geographiae Veteris* also by Bertius (1618–19); *Nova et accvrata Italiae Hodiernae descriptio* (1626; thirty-one maps and sixty-six views of towns).   His maps, though decorated with cartouches, arms and flamboyant lettering are generally restrained.

The brother-in-law of the older Jodocus Hondius, Pieter van den Keere, was born at Ghent in 1571, son of the celebrated typefounder Hendrik van den Keere.   The father died in 1580 and his mother remarried.   At the age of thirteen Pieter came to England.   As we have just seen, he later here met Jodocus Hondius, with whom he returned to Holland in 1593–4.   He worked as an engraver for both Hondius and Jansson, and died at Ghent in 1646.   Among van den Keere's works are five engravings and maps for John Norden's *Speculum Britanniae the first parte* (1593; *see* page 62; a map of Ireland published by Hondius in 1592; a series of small maps of English and Welsh counties (1599), based on those by Saxton (*see* pages 57*ff*); six maps of Scotland and five of Ireland; city panoramas and town plans; various world maps (1604, 1607, 1608, 1621); maps of Europe, Asia and America (1607–17), and of European countries (1604–29); an atlas of the Netherlands (1617); and the maps in John Speed's *Prospect of the Most Famous Parts of the World* (1646).   This by no means exhausts the list of works by this prolific engraver and publisher, whose signature is one of the most familiar of the period to collectors of maps.

Peter Plancius (1552–1622) of Drane-Outer in West Flanders, theologian, minister and former monk, is an important figure in the history of cartography. Indeed some say that in his own day he was considered to be second only to Mercator. His most important cartographic work was in the development of charts, but he also originated the characteristic Dutch world map printed on several sheets for mounting on cloth as a wall hanging. One such, the world in two hemispheres, he issued in eighteen sheets in 1592. Only one impression is now known. Altogether he made about eighty maps, but never published an atlas.

Plancius was among the founders of the Dutch East and West India Companies; thus he had a vested interest in the creation of good charts. He became official cartographer to the Dutch East India Company, and was also recognised as an authority on navigation especially on the route to the East Indies. He once toyed with the idea of a route via the north-east passage, but later advocated that round the Cape of Good Hope.

The van Langrens were an important family of Dutch cartographers of this period; they were also globe makers, astronomers and mathematicians. The family's founder was Jacobus Florentius van Langren; he was born in Utrecht but later lived in Amsterdam, of which city he was a freeman. His sons, Arnoldus Florentius and Henricus Florentius, and his grandsons, Michael Florentius and Jacob Florentius, were all cartographers. The family are particularly noted for their globes, the earliest of which probably dates from 1580, but Arnoldus and Henricus were also map makers (Plate 11). Their works include a map of Holland by Henricus (1594), a revised edition of an earlier one by Cornelis de Hooge. Henricus added to the original up-to-date data regarding changes made by drainage. Arnoldus produced a richly decorated world map in 1594, *Typus orbis terrarum*, on an oval projection. In 1596 the brothers produced another world map, a copy of one by Plancius, to which they added representations of the arctic areas. This, too, is richly decorated and includes allegorical female figures representing the four continents. Like most of their contemporaries, the Van Langrens inserted animals, monsters and ships on their maps, and used the conventional 'molehills' to represent relief.

In 1626 Arnoldus's son, Michael, made a map of the new canal from the Maas to the Rhine, *Fossae S. Mariae descriptio*. He also engraved maps for Blaeu. Michael was appointed mathematician to the King in 1628 and was paid a yearly salary of 240 livres for making various maps for him. Subsequently he was appointed king's cosmo-grapher in Flanders. Michael's elder brother, Jacob, also held an official position, that of cosmographer and engineer to his majesty, and was paid a hundred and fifty guilders for making various maps. He also engraved the earliest road book of England with maps: *A Direction for the English Traviller* (1635).[1]

Another great family of Dutch cartographers, among the greatest known, were the Blaeus (Plates 12 & 14). The family map-making firm was founded in 1596 by William Janszoon Blaeu (born at Alkmaar 1573; died at Amsterdam 1638); it issued maps, atlases, wall maps and globes. W. J. Blaeu was originally a globe and instrument maker. He had stayed with the great Danish astronomer, Tycho Brahe, on his islet Hven in the Sound, and owed much to Brahe for his development as a scientist.

Blaeu's first publications included sailing instructions, declination tables and charts. He was an ingenious mechanic and invented a new form of printing press. Writing of this in 1683, Joseph Moxon said (in *Mechanick Exercises*): 'The New-fashion'd Presses are used generally throughout the Low-Countries.' His workshop was extensive and included a typefoundry, paper manufactory, bindery, nine type-presses, six copper-plate presses, and proof-reading rooms, in addition to rooms for other departments of the printing processes. In 1633 he succeeded Hessel Gerritsz as cartographer to the Republic.

W. J. Blaeu published a number of atlases. He had, from 1605, published a number of maps of suitable size for an atlas, and in 1629 bought some copper plates that had become available on the death of the younger Jodocus Hondius. From combinations of these he published, beginning in 1630, a number of *Appendixes*, intended as supple-ments to the atlases of Ortelius and Mercator. However, the Mercator *Atlas* was considerably enlarged in 1633, making an appendix superfluous, and Blaeu, had he wanted to compete in the cartographic market, would have needed to publish an atlas of his own. This he did in 1634, with his *Novus Atlas* of one hundred and sixty-one maps on one hundred and fifty-six double sheets. Enlarged editions were published in 1635 (two volumes), and 1640 (three volumes). Fourth, fifth and sixth volumes were added in 1645, 1654 and 1655. As Blaeu died in 1638, he did not see the publica-tion of these three volumes, which were completed by his sons.

On his death, Willem was succeeded by his son Johann (1596–1673), noted for his work on the standardisation of sea charts. Johann's brother, Cornelis (d. 1642) worked

[1] *See also* page 63.

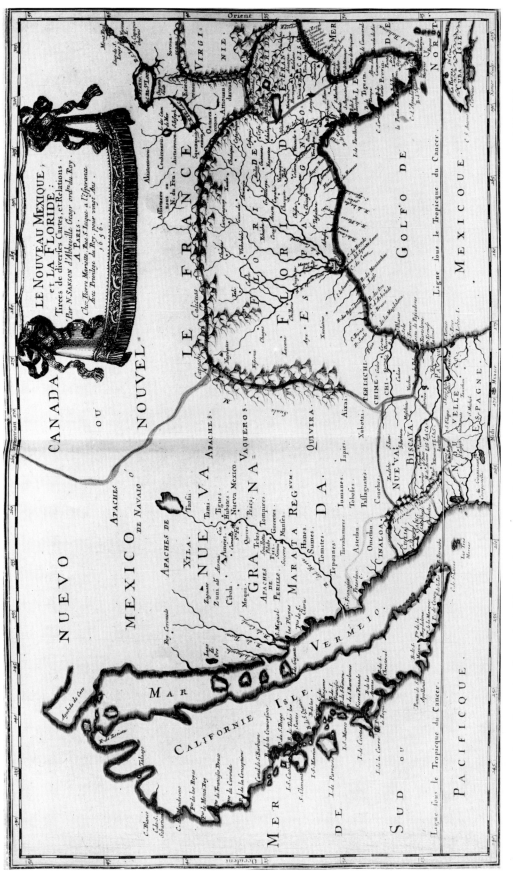

17. Map of Mexico, Florida and California from Sanson's *Cartes Générales*, 1656.

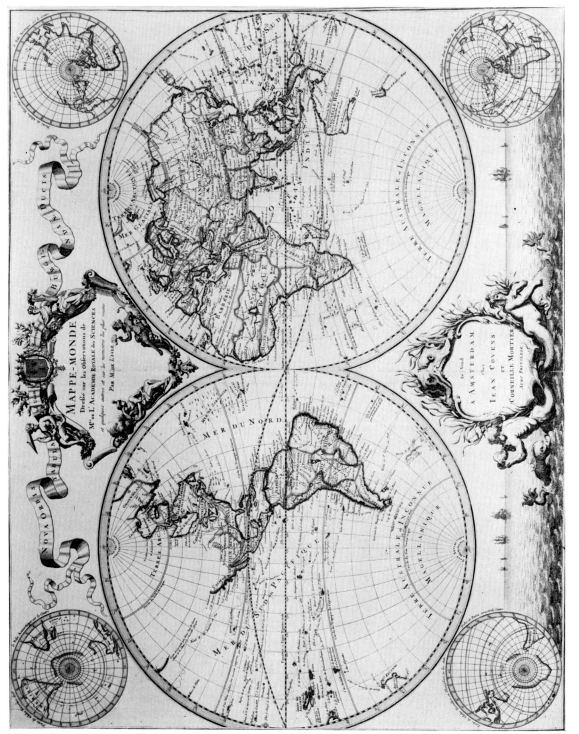

18. Map of the World in Hemispheres from De l'Isle's *Atlas Nouveau*, 1730.

British Museum.

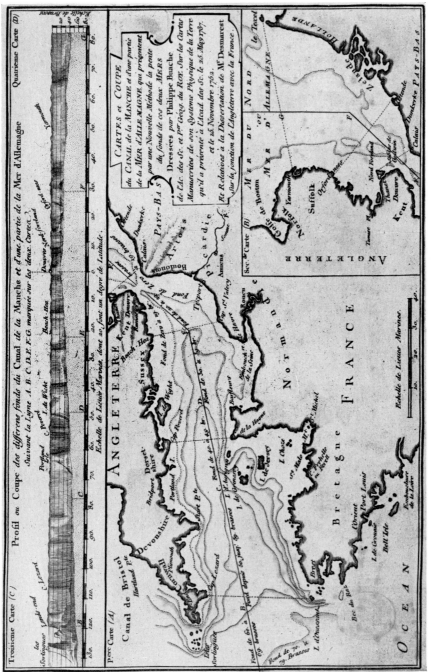

19. Map of the English Channel and German Sea by Philippe Buache, 1752.

20. Plan of Norwich from Cuningham's *Cosmographical Glasse*, 1559.   *University Library, Cambridge*

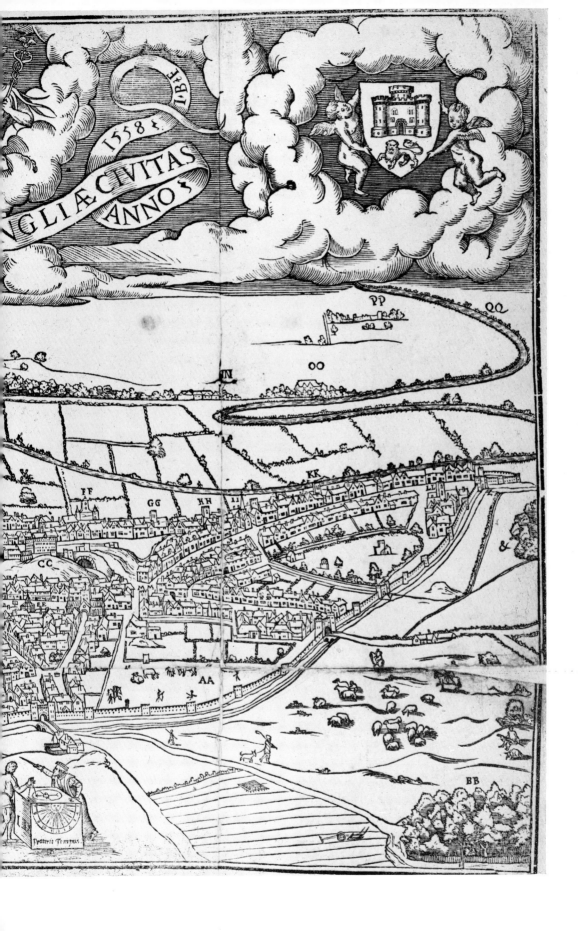

LIBER

1558.

...GLIÆ CIVITAS ANNO

N

PP   QQ

OO

KK

FF   GG   HH   II   LL

CC   &

EE

AA

BB

Præterit Tempus.

21. *Irelande* by Baptista Boazio, 1599.   *British Museum.*

Overleaf: 22. Plan of Cambridge from John Caius's *History of the University of Cambridge*, 1574. University Library, Cambridge.

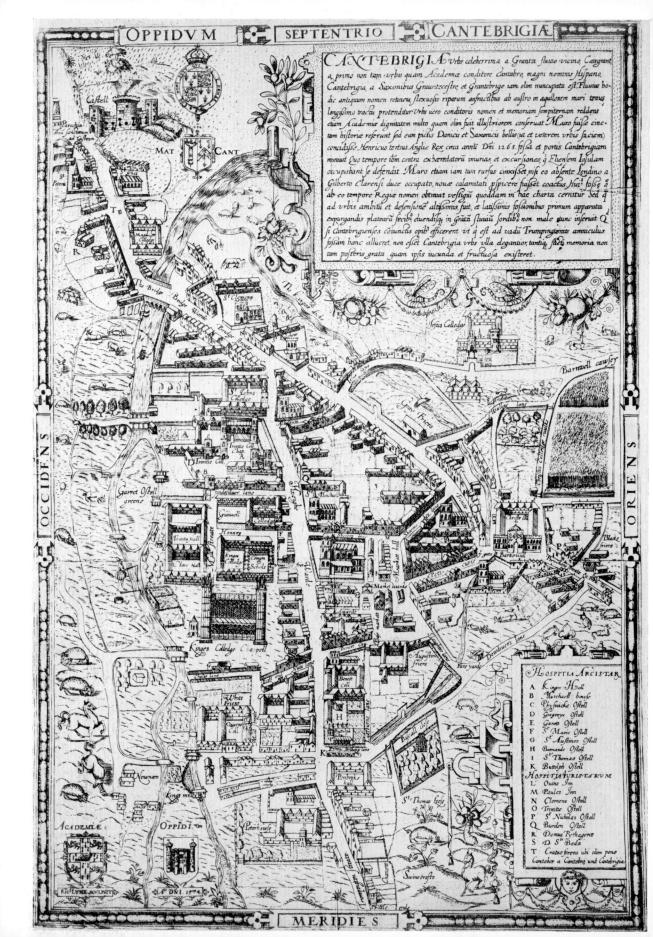

Castell

Parochia
omnium sanctorum
extra castrum

MAT CANT

St Petri

The Bridge

The Bridge Streete

St Clement

Magd College

Jesus Colledge

Barnwell cawsey

Sydney strete

Graye Fryers

Johns College

Trinitie coll.

D Ironike Coll.

Findesiluer lane

Garret Ostell greene

Griswell Colledge

Henney

S. Michaell

High strete

The School strete

Trinitie Hall

Christe Ch.

Blacke friers

Clare Hall

Barnwell

Petti cury

Kings Colledge Coappell

Market hill

Market howse

Augustine friers

Dowdivers lane

Fore yarde

CANTEBRIGIÆ urbs celeberrima a Granta fluuio vicino, Cairgrant a primo non tam vrbs quam Academæ conditore Cantabro, magni nominis Hispano, Cantebrigia, a Saxonibus Grauntzcestre et Grantebrige iam olm nuncupata est. Fluuius hodie antiquam nomen retineus flexuosis riparum anfractibus ab austro in aquilonem mari tenus longissimo tractu protenditur. Vrbs vero conditoris nomen et memoriam sempiternam reddens etiam Academiæ dignitatem multo quam olim fuit illustriorem conseruat. Muro fuisse cinctam historia referunt sed eum pictis Danicis et Saxonicis bellis (ut et veterem vrbis faciem) concidisse Henricus tertius Angliæ Rex circa annū Dñi. 1268. fossa et portis Cantebrigiam muniuit. Quo tempore ibi contra exhæredatorū inurias et excursiones, q Eluensem Insulam occupabant se defendit. Muro etiam iam tun rursus circuiSet, nisi eo absente Londino a Gilberto Clarensi duce occupato, nouæ calamitati prospicere fuisset coactus huiª fossæ q ab eo tempore Regiæ nomen obtinuit vestigiu quoddam in hac charta cernitur. Sed q ad vrbis ambitu et defensionē altissimi sunt, et latissimis fossionibus primum apparata expurgandis platearu secibª euendisq in grata fluuiū sordibª non male tunc inseruit Q si Cantebrigienses coniunctis opibª efficerent vt q est ad vadu Trumpington tu te amniculus fossam hanc allueret, non esset Cantebrigia vrbs vlla elegantior, tantiq factiq memoria non tam posteris grata quam ipsis iucunda et fructuosa existeret.

OCCIDENS

ORIENS

White Friers

Benett Coll.

Quenes Coll.

Peniboke Coll.

St Thomas leese

Pascall close

Newnam

Kings milns

Academia

OPPIDI

Peterhouse

Swine croste

Litle ende

RIC. LYNE SCVLPSIT

Aº DÑI. 1574.

Hospitia Archistar.
A Kinges Hall
B Michaell house
C Physicke Ostell
D Gregoryn Ostell
E Garett Ostell
F St Marie Ostell
G St Austenes Ostell
H Barnards Ostell
I St Thomas Ostell
K Buttolph Ostell
Hospitia Iuristarum
L Ouins Inn
M Paules Inn
N Clemens Ostell
O Trinitie Ostell
P St Nicholas Ostell
Q Burden Ostell
R Domus Pythagore
S D St Beda
T Crates ferrea vbi olim pono
Cantebr a Cantebrig vnd Contebrigia.

with him in the firm. They produced a world map in 1606, of which only one copy has survived, but their most important works were their completion of the *Atlas Novus* and *Atlas Maior* (1662; eleven volumes), which went through a number of editions. The *Theatrum* contained the first national atlas of Scotland; it was in one volume. It also included an atlas of the counties of England and Wales, based on the maps of Saxton and Speed (*see* pages 57*ff* and 66*ff*). Johann Blaeu relates, in the Preface to the Scotch atlas, how the maps it contained originated:

> Timotheus Pont, a native of Scotland, travelled far and wide over the whole of that Realm, viewed it with attention, and laid it down in some maps, although rather roughly, because in the beginning all things are imperfect. Ioannes Scot, lord of Scotis-Tarvet, and recently still Director of the Chancellery of Scotland, has with wonderful industry collected these and other maps, guarded them like a precious treasure against shipwreck, and sent them over to me, but much torn and deformed. I have brought the same in order, and sometimes divided a single map, drawn in a confused and incorrect way, in several parts, and in particular I have kept each county separate. After that Robert and James Gordon have given this work the finishing touches, as the saying goes, and have added thereto, apart from the corrections in the maps of Timotheus Pont, a few of their own, together with some descriptions, as well by themselves as by others. A certain Orcadian, whom I previously employed in my printing-office to correct the errors of the typesetters, I have caused to describe the Orcades and the Shetland or Hethland Islands. The remaining descriptions have been adopted from Buchanan and Camden. But the latter author being an Englishman, and, according to his own words, not so experienced in the Scottish writings, his descriptions have in many places been actually corrected by the abovementioned lord of Scotis-Tarvet.

There is some complication in the names used by W. J. Blaeu. Until 1617 Willem used his patronym, Janszoon, signing his work Willems Jans Zoon, or, in Latin, Guilielmus Janssonius; sometimes it was Englished as William Johnson. Sometimes he used his christian names only, combined with the name of his birthplace; Willem Jansz. Alcmar, or Guilielmus Janssonius Alcmarianus. Some care is therefore needed in identifying his maps, and particular care must be taken not to confuse him with his contemporary, Jan Jansson (*see* page 43). Later he signed his name either as G. Blaeu or with his christian name expanded, Guilielmus Blaeu. Sometimes, from 1621, his surname is spelled Blaeuw.

The English diarist, John Evelyn, visited Blaeu's establishment in 1641, as recorded in his Diary. 'I went,' he writes, 'to Hundius's shop to buy some maps, greatly pleased with the designs of that indefatigable person. Mr Bleaw, the setter forth of the Atlas's and other works of that kind, is worthy seeing.'

A younger contemporary of W. J. Blaeu was Hessel Gerritsz, who was born in

1580–1 in the village of Assum.   He served, from 1599, an apprenticeship as an engraver under Blaeu, but for how long is not known.   In 1607 he married, after which he started his own business, but still continued to engrave for Blaeu.   It is known that by 1610 he had his own printing office at Amsterdam.   In 1613 he engraved for Blaeu a map of Lithuania, *Magni ducatus Lithuania*, and in the same year published a map of Russia, which appeared in a second edition in 1614.   This showed some advance on earlier maps of the area, though it was still full of inaccuracies.   Gerritsz also made maps of Spain, Italy, and Scandinavia, the last being published posthumously.   He made two sea charts, one of Ireland and one in manuscript form of the coasts of France.

By 1612 Hessel Gerritsz was working for the Admiralty at Amsterdam.   In the same year was published his book *Beschryvinghe Vander Samoyeden Landt in Tartarien*. It dealt with such apparently diverse subjects as Isaac Massa's[1] descriptions of Siberia, the roads and rivers from Muscovy eastward, and a map of the world in two hemispheres showing Hudson's discoveries, with an account of Hudson's fourth voyage in 1610.   It is of considerable rarity.   In 1613 he published another book *Histoire du Pays nommé Spitsberghe*.

In 1617 Gerritsz was appointed cartographer to the Dutch East India Company.   He was also chief of the Hydrographical Office of the Company, and as such all maps, drawings and publications sent to the Company went through his hands; he was thus able to check and correct any information he used.   During the years 1617–22 he made charts of the Indian archipelago and of Australia, Sumatra, the south-east and east Asiatic Islands, the Indian Ocean and the Pacific Ocean.   But of these, only the first was engraved.

Gerritsz was no mere academic geographer, for in 1628 he undertook a voyage to the New World.   He sailed on 25 October in that year on a ship called the *Zutphen*, which belonged to the fleet of Adriaen Janszoon Pater.   The voyage took him to the Azores, the Cape Verde Islands, Brazil, the Caribbean Islands and Cuba, returning home almost exactly one year later.   As a result of these voyages Gerritsz compiled two manuscript *routiers*, giving many coast profiles.   He also made a number of manuscript and engraved charts of the South and Central American waters.   Gerritsz died in 1632 and was succeeded as cartographer to the East India Company by W. J. Blaeu.

Gerritsz was a man of great versatility—author, publisher, printer, bookbinder,

[1] Isaac Massa (1586–1643) was born at Haarlem; his parents came from Antwerp. He travelled to Russia and brought back descriptions and maps of parts of the country, which he obtained at considerable risk to himself and his Russian collaborators, who would have been considered traitors for even collecting such information.

bookseller, geographer, as well as a cartographer with few equals in the seventeenth century. His maps were used as models for a long time after his death. On the whole they are, for their period, very restrained in design, and without decoration.

Jan Jansson, Blaeu's rival, whom we have already briefly mentioned, was born at Arnheim in 1596 and died at Amsterdam in 1664. His work is generally not so good as Blaeu's; perhaps this is because he tried to publish much of it too quickly to forestall Blaeu's publications. Decoratively it is first class, combining restraint with a tactful use of decorative cartouches, coats of arms, figures, putti and swash lettering (Plate 15). Jansson also published an edition of Ptolemy's *Geography* in 1617; and a *Théâtre du Monde* which was published in various languages and editions between 1639 and 1661, and varied, according to the edition, from three to eleven volumes. He also published globes.

The title page to Volume IV of the 1659 edition of Jansson's atlas, which is devoted to the British Isles, is a magnificent specimen of baroque book illustration. The main part of the design is a structure something between a Dutch cabinet and a mammoth fair organ. It is decorated with arms (the royal arms of England are at the top) and figures: Roman, Saxon, Dane, Norman, and Britannicus, the last-named a warlike Ancient Briton wearing only a chain and cloak, holding a shield and brandishing a spear. A panel at the top of the design displays the title; and a cartouche at the base, with the publisher's name and the date and place of publication, is held by kneeling putti. Above all this, in the sky, more putti hold shields, and sunrays shine down from a break in the clouds (Plate 16).

Jansson was succeeded in business by Peter Schenk (1645–1715), who worked in partnership with Gerard Valck (1650/1–1720). They republished Jansson's atlas and issued several works of their own, including *Hecatompolis* (1702), and *Le Théâtre de Mars* (1706). Later Gerard Valck's son Leonard worked with them.

Other Netherlandish map publishers who should be briefly mentioned are the Visscher family, founded by Claes Jansz (1587–1637) who worked first for Hondius and then set up his own business, in which he was succeeded by his son (1618–*circa* 1679) and grandson (1649–1709), each having the same name as himself. There were also Karel Allard (1648–1706); and the Danckerts family, of which there were several members working until the end of the first quarter of the eighteenth century. In the eighteenth century there were the firm of Covens and Mortier, Pieter van der Aa of Leyden (*fl.* 1713–30) and Isaac Tirion (d. 1769), publisher of several atlases. Important work was made and published by each of these cartographers, but a special word must be said for van der Aa's *tour de force* in publishing *La Galerie Agréable du Monde*, an atlas of sixty-six volumes and some three thousand plates.   But the greatest era of Dutch cartography had passed by the end of the seventeenth century, and by the eighteenth century Dutch maps had become imitative and decadent.

BIBLIOGRAPHY

BAGROW, LEO: *History of Cartography*.   1964.

GERNEZ, D.: 'Lucas Janszoon Waghenaer'.   *Mariner's Mirror*, Vol. XXIII.   1937.

HEAWOOD, E.: *The Map of the world on Mercator's Projection by Jodocus Hondius 1608*.   1927.

—— 'A Masterpiece of John Blaeu'.   *Geographical Journal*, Vol. LV.   1920.

—— 'An unrecorded Blaeu map of *c*. 1618'.   *Geographical Journal*, Vol. CII.   1943.

KEUNING, JOHANNES: 'Blaeu's *Atlas*'.   *Imago Mundi*, Vol. XIV, pp. 74–89.   1959.

—— 'Cornelis Anthonisz'.   *Imago Mundi*, Vol. VII, pp. 51–65.   1950.

—— 'Dutch Cartography in the XVIth Century'.   *Imago Mundi*, Vol. IX.   1952.

—— 'Hessel Gerritsz'.   *Imago Mundi*, Vol. VI, pp. 49–66.   1949.

—— 'The History of an atlas.   Mercator-Hondius'.   *Imago Mundi*, Vol. IV, pp. 37–62. 1947.

—— 'Isaac Massa, 1586–1643'.   *Imago Mundi*, Vol. X, pp. 65–79.   1953.

—— 'Jodocus Hondius Jr.'   *Imago Mundi*, Vol. IV.   1947.

—— 'The Novus Atlas of Joannes Janssonius'.   *Imago Mundi*, Vol. VIII.   1951.

—— 'Pieter van den Keere (Petrus Kaerius), 1571–1646 (?)'   *Imago Mundi*, Vol. XV, pp. 66–72.   1960.

—— 'XVIth century cartography in the Netherlands'.   *Imago Mundi*.   Vol. IX.   1952.

—— 'The Van Langren Family'.   *Imago Mundi*, Vol. XIII, pp. 101–9.   1956.

KOEMAN, C.: *The History of Lucas Janszoon Waghenaer and his 'Spieghel der Zeevaerdt'*.   1964.

LYNAM, E.: 'Lucas Waghenaer's Thresoor der Zeevaert'.   *British Museum Quarterly*, Vol. XIII.   1939.

STEVENSON, E. L.: *Willem Janszoon Blaeu 1571–1638*.   1914.

TOOLEY, R. V.: *Maps and map-makers*.   1962.

CHAPTER FOUR

France

WE must now turn to France. Before the seventeenth century French cartography had been unimpressive, despite some outstanding practitioners here and there. In addition to those already mentioned in Chapter Two, there was Charles de l'Escluse (Carolus Clusius; 1525–1609) of Arras, who was well known as a botanist—he grew the first horse-chestnut tree in Vienna in 1576 from seeds he had found at Constantinople. He made a map of Spain of which only one impression is known. Among other cartographers were Jean Jolivet (*fl.* 1545–60), who made woodcut maps of France and Picardy, and an engraved map of Berry; Gilles Boileau de Bouillon (*fl.* 1551–63), also known by the pseudonyms of Darinel and Pasteur des Amadis, who made maps of Belgium, Savoy and the Roman Campagna; Gabriel Symeone (1509–75), maker of the map *La Limagna d'Overnia*; André Thevet (1502–90), author of *La Cosmographie Universelle* and maker of several maps, including one of the world in the shape of a lily—doubtless a compliment to the rulers of his country. There are others also, but the works of all of these early French cartographers are of the greatest rarity.

There was an important development at Tours in 1594 when Maurice Bouguereau issued *Le Théâtre Françoys*, the first French national atlas. It contained, according to the copy, fourteen to sixteen line engraved provincial maps of France, and from one to three general maps. Some of the maps had been issued previously by Mercator, Ortelius and de Jode, but there were new maps of Blasois by Jean du Temporal, Limousin by Jean du Fayen and Touraine by Isaac Françoys. The only map by Bouguereau himself was that of Poitou. The *Théâtre Françoys* is a very rare work. It was reissued by Jean le Clerc under the new title of *Théâtre Géographique du Royaume de France*. Several editions were issued under the new title between 1617 and 1631, by which time the number of maps had grown to fifty-two. The plates were acquired before 1642 by the publisher, who issued it with seventy-five maps in 1642, under the title of *Théâtre des Gaules*.   Another atlas was issued in 1634 and again in 1637 by Melchoir Tavernier (1544–1641), with the same title as the Le Clerc edition of the foregoing atlas—*Théâtre Géographique du Royaume de France*.   The content of copies varies from eighty to ninety-five maps.

The Taverniers, of Flemish origin, were a large family of map-publishers and cartographers.   Melchoir had three sons of whom one, Melchoir Junior (1594–1665), was a cartographer.   The others, Jean Baptiste and Daniel, were travellers.   Melchoir's brother, Gabriel, had a map-selling business at Antwerp.

One of the greatest families in seventeenth-century French cartography were the Sansons who made maps for over a hundred years (Plate 17).   They began with Nicolas Sanson who was born in 1600 at Abbeville and died at Paris in 1667; he became 'géographe ordinaire du roi'.

Nicolas Sanson published some three hundred maps, beginning in 1629 with a six-sheet map of ancient Gaul.   Though it seems that at first his main intention was to publish single maps, Sanson produced a number of atlases, including his well-known *Cartes générales de toutes les parties du monde*, which in its first edition contained one hundred and thirteen maps.   A subsequent edition published after his death by his sons Guillaume (d. 1703) and Adrien (d. 1708) contained only one hundred and two maps.

Various assemblers of atlases used Sanson maps for their compilations.   Among such compilers was B. J. Briot.   In the main, craftsmen from Picardy and the Low Countries engraved Sanson's maps.

Other works published by the Sansons include octavo atlases of the four continents: *L'Europe en plusiers cartes nouvelles* (1648 and 1651), *L'Asie* (1652, 1653 and 1658), *L'Afrique* (1656 and 1660), and *L'Amérique* (1656, 1657, 1662 and 1676).

Pierre du Val (1619–83) Sanson's son-in-law, published a series of important works, among them the small *Cartes de géographie* (1662; forty maps), *Cartes et tables de géographie des provinces eschués* (1667; ten maps) and *Le Monde ou la Géographie Universelle* (1670; eighty-four maps); each of these works was 12 mo. (5 to $5\frac{1}{2}$ by $7\frac{3}{8}$ to $7\frac{1}{2}$ inches).

The Sanson business was bought from Sanson's sons by Charles Hubert Alexis Jaillot (b. Avignon 1640; d. Paris 1712).   Jaillot began life as a sculptor, but turned to cartography when he married a daughter of the cartographer and map colourist, Nicolas Berey (*fl.* 1650).   He had worked with the Sanson sons before taking over their business.   When he acquired their plates he had them re-engraved on a larger scale; he also made maps and plans himself.

Jaillot's publications included the *Atlas Nouveau* of 1681, which passed through several editions and became, in 1695, the *Atlas François*, under which title it again passed through several editions.   In 1693 Jaillot issued a sea atlas, *Le Neptune Français ou Atlas Nouveau des Cartes Marines*.   The Jaillot business was run by his descendants until 1780.   They included Bernard Antoine, Bernard Jean Hyacinthe, and Chauvigne Jaillot.

Decoratively, Jaillot's maps reached a level that has never been surpassed.   To take one instance, his map of Palestine of 1691 has a cartouche containing explanatory notes and scales.   The lower part of this is formed of two cornucopias overflowing with fruit and leaves.   The upper part is a decorative representation of the Garden of Eden. The tree of knowledge, complete with entwined serpent, is at the apex and Adam and Eve stand on either side, naked but for their aprons of fig leaves; Eve is pointing at the tree.   Around Adam and Eve are grouped beasts from the Garden: elephant, camel, lion and lioness, bull, horse and peacock.   Such a bare description gives little idea of the baroque magnificence of the work.

Nicolas de Fer (1646–1720) was an engraver and geographer, and the creator of a wide range of maps, particularly notable for their decorations.   His works include the magnificently titled *La France triomphante sous le régne de Louis le Grand* (six sheets; 1693), *Plusieurs cartes de France avec les routes et le plan des principales villes* (1698), *Atlas royal* (eighty-six maps; 1699–1702), *Atlas curieux* (one hundred and twenty-two maps and plates; 1700–3), and a series of provincial maps of France.

The cartographer Nicolas Tassin (*fl.* 1633–55) is noted particularly for his *Plans et Profilz de toutes les principales villes . . . de France* (one hundred maps), which ran through three editions: 1634, 1636 and 1638.   But he produced other works, among them *Descriptions de tous les Cantons, Villes, Bourgs des pays de Suisses* (1635), *Cartes Générales de la Géographie royalle* (1655; eighty-four maps), and *Cartes générales de toutes les costes de France* (1634).

There was an upsurge of science and technology in France during the reign of Louis XIV.   At the same time there was a great call for maps and charts as a background and aid for French expansion and ambitions.   It was quickly appreciated that if they were to be of value, they must be based on accurate observations and data.   There was therefore an attempt to make French cartography as precise and scientific as possible.

A leading figure in this movement was the Italian astronomer and geographer Giovanni Domenico Cassini (1625–1712), who was born at Perinaldo.   He became professor of astronomy at Bologna, and was later invited by Louis XIV to come to

France.   Thereafter, in 1671, he became professor of astronomy at the Collège de France, and from 1669, director of the Royal Observatory at Paris.   He made several innovations in cartographic methods, including the use of the movements of the moons of the planet Jupiter to determine longitude.   It was as a direct result of Cassini's work that a new map of France was made and a new survey of its coasts carried out by Philippe de la Hire and the Abbé Picard.   One result of this survey was a recession of the coastline of France on the map.   This caused Louis XIV to remark that the survey had lost him more territory than a disastrous war.

Another Cassini, Jacques, and his son, César François, began in 1733 to prepare a map of France based on triangulation of the whole country.   Triangulation is a method by which the earth's surface may be measured by means of the geometric laws of triangles; in this way a position or location may be determined by relating it to two fixed points a known distance apart.   C. F. Cassini's survey of France by triangulation received official support in 1747, but this was withdrawn soon after, and Cassini had to borrow money to complete his work.   By the time he died in 1784 he had completed maps of the whole of France with the exception of Brittany, on a scale of approximately one and a third miles to the inch.   After the Revolution, the state took over responsibility for the survey, which was completed in 1818.

In its completed form the Cassini map looks almost like a modern ordnance survey map, though fifty years were to elapse from its commencement before the British Ordnance Survey was founded.   Symbols are used extensively for topographical features, but relief is poorly shown.   Colour is used for various features, such as red for main roads, and yellow for secondary roads.   Some decoration is still used, for eighteenth-century French cartographers took this aspect of their subject seriously. Some of the famous artists of the day, such as Cochin and Boucher, decorated maps; which meant that the most scientific maps so far seen were presented in the most artistic and elegant manner imaginable—a rare combination.

One of the most famous figures in eighteenth-century French cartography was Guillaume de l'Isle or Delisle (1675–1726), son of Claude de l'Isle (1644–1720), a teacher of geography and other subjects (Plate 18).   Guillaume had been a pupil of G. D. Cassini, and it was he above all who popularised the Cassinis' work.   Guillaume began his work as a map-publisher in 1700 and thereafter became a leading cartographer. De l'Isle was elected a member of the Académie in 1702, and in 1718 became *Premier Géographe du Roi*.   Among his foreign patrons was Peter the Great of Russia.

One of de l'Isle's earlier productions was *Mappe-Monde Dressée sur les Observations de*

Clemens et Regni moderatrix uſta Britäm
Hac forma inſigni conſpicienda nitet.

Triſtia dum gentes circum omnes bella fatigant,
Cæciſ errores toto graſſantur in orbe.
An. Dñi Pace beas longa, vera et pietate Britannos: 1579
Iuſticia moderans miti ſapienter habenas.
Chara domi, celebriſ foris, longæuaſ regnü
Hic teneas, regno tandem fruitura perenni.

23. Engraved frontispiece from Saxton's Atlas, 1579.   *British Museum.*

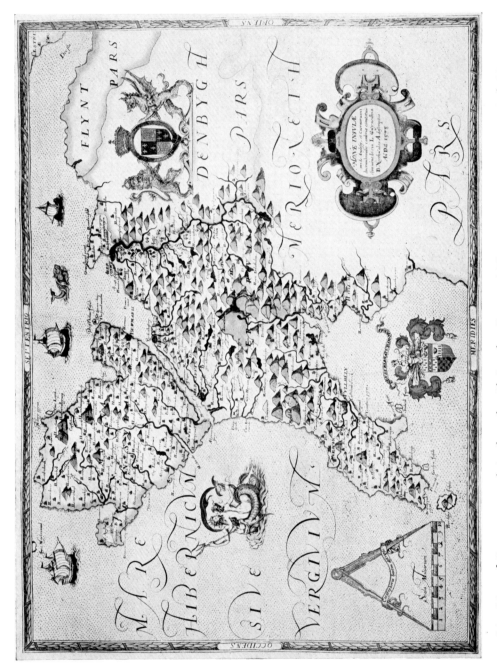

24. Map of Caernarvon and Anglesey by Christopher Saxton, 1579.

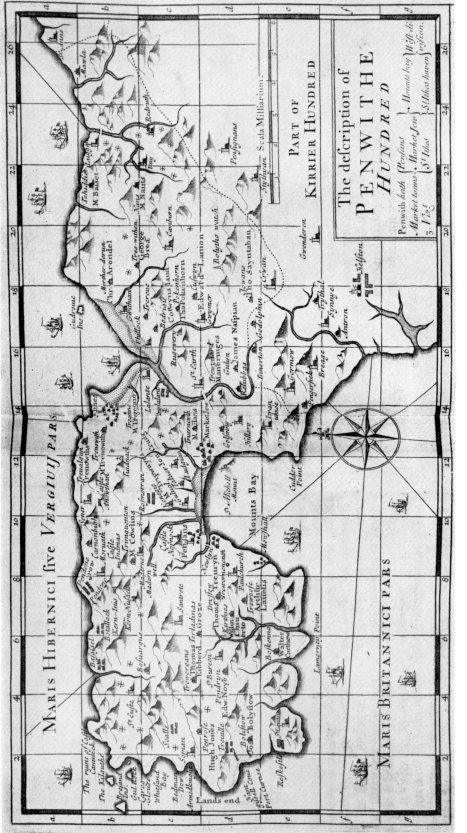

25. John Norden: Map from *The Description of Cornwall*, 1728. *Private Collection.*

26.   Map of Norfolk by Christopher Saxton, 1579.   *British Museum.*

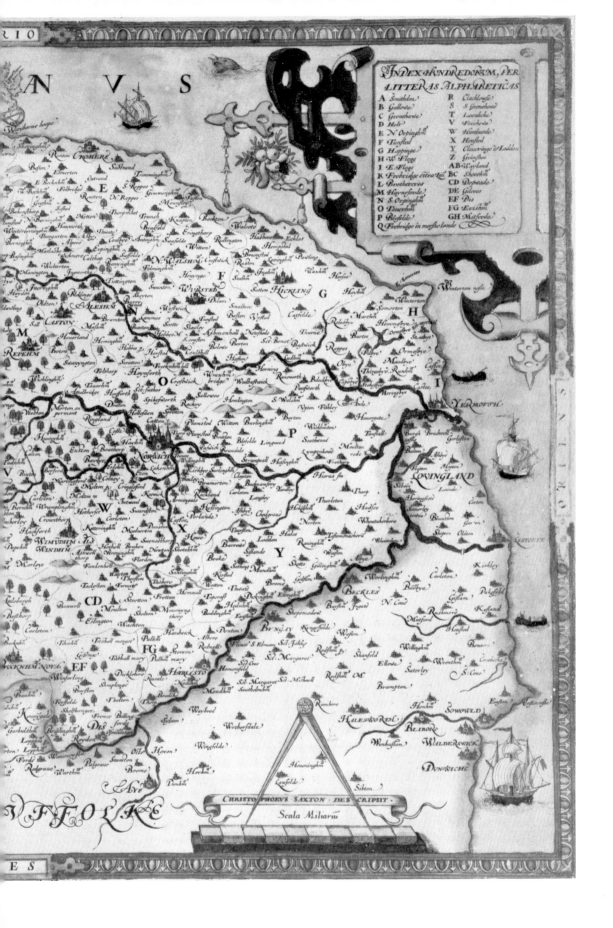

Somerset.

	Brightstoll	Bathe	Welles	Shepton	Bruton	Somerton	Ilchester	Glastonbury	Bridge-water	Taunton	Charde	Euell	Wellington	Wilcombe	Dunster	Crooke-horne	Froome	Wincaunton	Hunspill	Lamporte	Miluerton	Duluerton	Whatchet	Pensforde	Ilmister
Ax-bridge.	12	16	8	11	16	14	17	9	13	18	23	21	24	23	28	23	18	20	10	15	23	32	30	22	11
Ilmister.	32	32	19	20	21	10	12	15	11	7	4	12	12	15	23	5	28	21	15	6	13	23	25	20	29
Pensforde.	5	7	11	11	16	20	22	15	22	29	32	27	35	34	36	30	12	19	19	31	32	42	39	32	
Watchet.	34	36	25	27	31	22	25	12	20	12	20	27	12	7	5	24	36	33	14	19	11	11	7		
Mynehead.	46	43	31	34	38	28	30	29	18	17	25	34	15	11	3	27	42	38	20	24	10	11			
Duluerton.	43	46	34	36	40	30	31	41	20	16	22	33	11	8	28	45	41	23	24	11					
Miluerton.	34	38	25	27	30	19	21	22	11	6	14	24	3	13	18	35	31	14	15						
Lamporte.	27	26	14	15	15	5	6	8	8	11	12	11	15	17	22	8	22	16	12						
Hunspill.	20	24	13	16	20	13	16	11	5	12	18	20	17	16	18	19	24	23							
Wincaunton.	24	22	12	9	3	12	10	13	23	25	23	12	30	33	37	19	10								
Froome.	16	8	11	8	18	18	18	15	25	30	31	20	36	37	40	27									
Crooke-horne.	33	22	20	20	19	10	8	15	15	12	6	8	16	20	27										
Dunster.	38	42	30	31	35	26	27	26	16	15	23	31	13	8											
Wilcombe.	36	40	38	29	31	22	23	23	13	8	15	25	5												
Wellington.	37	39	26	28	30	19	21	22	13	5	12	23													
Euell.	29	27	15	14	12	7	4	12	17	18	14														
Charde.	35	36	22	23	27	13	11	18	14	9															
Taunton.	31	33	20	22	24	15	15	16	7																
Bridgewater.	24	27	14	16	20	11	14	13																	
Glastonbury.	19	18	4	6	10	6	8																		
Ilchester.	26	24	12	12	11	4																			
Somerton.	24	22	10	10	11																				
Bruton.	20	16	8	6																					
Shepton.	16	13	4																						
Welles.	15	14																							
Bathe.	9																								

The vfe of this Table.

THe Townes or places betweene which you defire to know, the diftance you may finde in the names of the Townes in the vpper part and in the fide, and bring them in a fquare as the lines will guide you : and in the fquare you fhall finde the figures which declare the diftance of the miles.

And if you finde any place in the fide which will not extend to make a fquare with that aboue, then feeking that aboue which will not extend to make a fquare, and fee that in the vpper, and the other in thefide, and it will fhowe you the diftance. It is familiar and eafie.

Beare with defectes, the vfe is neceffarie.

Inuented by IOHN NORDEN.

27. John Norden: Distance Tables from *England: An intended Guyde*, 1625. *British Museum.*

28. Thomas Seckford's portrait on a copper token of Woodbridge, Suffolk, 1796. *Private Collection.*

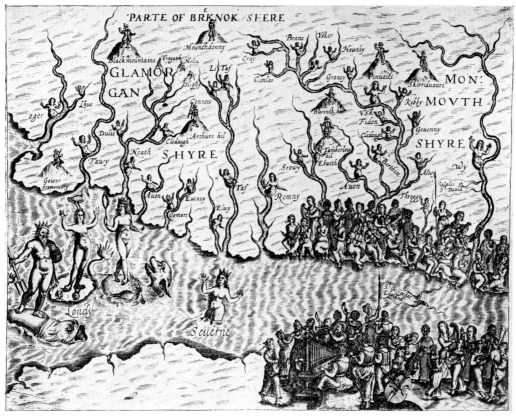

29. Map of St. George's Channel from Michael Drayton's *Poly-Olbion*, 1622. Engraved by William Hole. *Private Collection.*

30. Map of Oxfordshire, Berkshire and Buckinghamshire from Michael Drayton's *Poly-Olbion*, 1622. Engraved by William Hole. *Private Collection.*

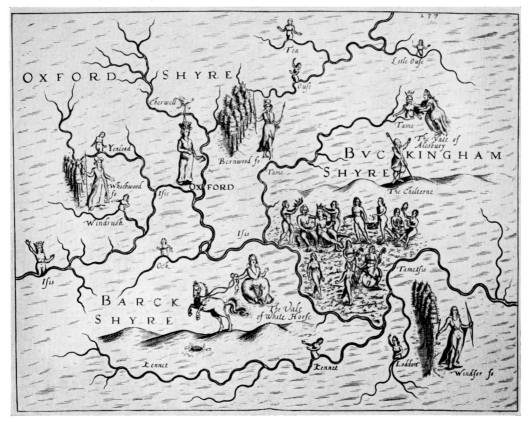

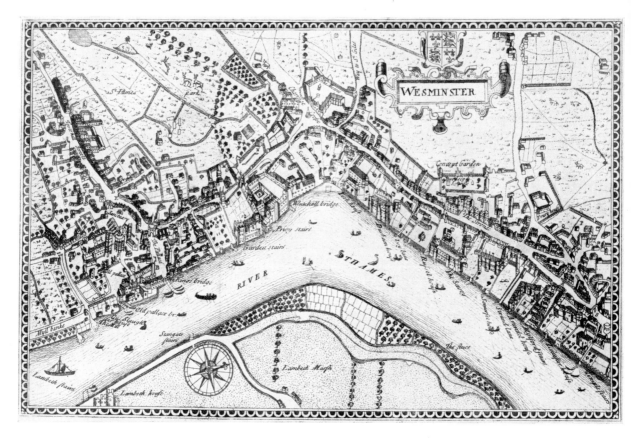

31.   Plan of Westminster.   From Norden's *Description of Middlesex and Hartfordshire,*
      1723 edition.   Engraved by Senex.   *Private Collection.*

Mrs de l'Académie Royale des Sciences; it was in two hemispheres, and in its outlines of the continents was remarkably accurate (Plate 18). Moreover, de l'Isle had not indulged in guesswork to fill in blanks. Where nothing was known he simply left a space. Several subsequent editions of the map were published, with improvements and emendations. Not for nothing is de l'Isle known as the initiator of the reformation of cartography. He issued a total of about a hundred maps.

De l'Isle's improvements to the standards of accuracy in cartography were continued by the scholar-cartographer Jean-Baptiste Bourguignon d'Anville (1697–1782). D'Anville's great collection of between ten and twelve thousand cartographical items is now in the Bibliothèque Nationale at Paris. His approach to the subject was essentially scholarly, and consisted of the correlation of material. He made an especial contribution to Asiatic cartography and was among the first to study the works of oriental writers for source material. He published a survey of the provinces of China for the Society of Jesus, which enabled him to make a map of the Chinese Empire; it occupied sixty-six sheets, and had a geographical description of the Empire by J. B. du Halde. This work was later published also in Holland, under the title of *Nouvel Atlas de la Chine de la Tartarie Chinoise* (1737), and also in England (1738–41; Plate 54).[1]

Other maps by d'Anville included a general map of the world in two hemispheres (1761), a map of India (1752), and maps of North America, South America, Africa, Asia, and Europe, published between 1746 and 1760. In these he followed and sur-passed de l'Isle in removing imaginative topography, which meant that, generally speaking, detail remained only in coastal areas. He issued an *Atlas antiquus major* in 1768 and an *Atlas général* in 1780. Artistically d'Anville's maps are first class and show a high standard of draughtsmanship and engraving.

D'Anville's successor was his son-in-law Philippe Buache (1700–73), who made valuable contributions to the representation of relief on maps. As we have seen, this was once done by drawing views of little molehill-shaped hills. Later, but still well before the time of Buache, a Swiss cartographer, Hans Konrad Gyger (1599–1674), used a method of shading. While this was an improvement in that relief was shown in plan instead of in elevation, it could not be used effectively to show differences in height. From this probably was derived the method of hachuring to denote relief. Hachuring consists of lines drawn close together, running in the direction of the slope; the steeper the slope the heavier the hachuring. Again, while this was a definite step forward it

[1] *See also* p. 102.

D

was not accurate in detail. Buache's method was that generally used today, that is drawing lines through points of constant height (Plate 19). The idea was probably adapted from that used in 1729 by the Dutch hydrographer, N. S. Cruquius, for indicating equal ocean depths, although it may have originated from a chart of magnetic variations published by the English astronomer Edmund Halley (1656–1742) in 1701, on which he used isogonic lines: lines, that is, joining points of the surface of the earth at which magnetic declination is the same. In some modern maps, those published by Michelin for example, contour lines and hachuring are combined, sometimes with the addition of colour, to show relief.

Among works published by Buache are *Considérations géographiques* (1753–4), *Cartes et tables de la géographie physique* (1754), *Atlas géographique et universelle* (1762), and *Atlas géographique de quatres parties du monde* (from 1769). In his own times he was noted chiefly for his marine maps. After his death his plates, including those he had taken over from de l'Isle, passed to the publisher J. A. Dezauche.

Gilles (1668–1766) and Didier Robert de Vaugondy (1723–86) were members of another family of eighteenth-century French cartographers. In 1757 they published their *Atlas Universal* of one hundred and eight highly decorative maps, based on Sanson maps which they had acquired from Pierre Moulard Sanson, grandson of Nicolas Sanson. The price was 126 livres unbound. Madame de Pompadour was a subscriber to the publication.

Another eighteenth-century French cartographic publisher was Roch-Joseph Julien (*fl.* 1751–6) who in 1751 published an *Atlas géographique et militaire de la France*, followed by *Atlas topographique et militaire* (1758), *Nouveau théâtre de la guerre* (1758) and *Le théâtre du monde* (2 vols., 1768). The last-named includes maps taken from a wide variety of sources, among them the maps of d'Anville, de Vaugondy, de l'Isle, Homann, Jaillot, Sanson, Buache, Halley, de Wit and many others. Julien claimed that he kept four thousand maps in stock.

Jacques Nicolas Bellin (1703–72), cartographer and publisher, published a number of works including an *Atlas maritime* (1751), *Neptune français* (1753) and *Hydrographie française* (1756–65). G. L. Le Rouge (*fl.* 1741–79) also issued a series of maps and atlases, and especially plans of ports and fortifications. His works include *Théâtre de la guerre en Allemagne* (1741), *Recueil des Plans de l'Amérique* (1755), *Recueil des côtes maritimes de France* (1757), *Topographie des chemins de l'Angleterre* (1760), *Pilot Amériquaine Septentrional* (1778–9).

French supremacy in cartography dwindled as the eighteenth century waned. The outbreak of the Revolution and other internal difficulties accelerated this process. Thereafter England took the lead as her own maritime power began to climb towards its zenith.

BIBLIOGRAPHY

BAGROW, LEO: *History of Cartography.* 1964.

BERTHAUT, H. M.: *La Carte en France 1750–1898.* 1898–9.

—— *Les Ingénieurs géographes militaires 1624–1831.* 1902.

BROWN, L. A.: *Jean Dominique Cassini and his world map of 1696.* 1941.

DAINVILLE, FRANÇOIS DE: 'Jean Jolivet's "Description des Gaules"'. *Imago Mundi,* Vol. XVIII. 1964.

FORDHAM, SIR H. G.: 'The Cartography of the Provinces of France, 1570–1757'. *Studies in Carto-Bibliography.* 1914.

—— *Note on a series of early French atlases, 1594–1637.* 1921.

—— *Some notable surveyors and map-makers of the sixteenth, seventeenth and eighteenth centuries.* 1929.

—— *Studies in Carto-bibliography, British and French.* 1914.

TOOLEY, R. V.: *Maps and map-makers.* 1962.

WATERS, D.W.: *The Rutters of the Sea. The Sailing Directions of Pierre Garcie. A study of the first English and French printed sailing directions . . .* 1967.

CHAPTER FIVE

Great Britain

S o far as printed maps are concerned, the real foundations of English cartography were laid in the sixteenth century. Early in that century many local plans were made, largely because of the demands of those who had acquired monastic properties at the Dissolution in 1536 and the Suppression in 1539. Among the first to have maps made of newly acquired estates was the Earl of Pembroke. There were, too, many disputes caused by enclosure of common land, which created a demand for good land-surveyors and gave rise to such books as Valentine Leigh's *The Most Profitable and Commendable Science of Lands, Tenements, Hereditaments* (1562 and subsequent editions), and the itinerary of England and Wales by John Leland (1506?–52), which was circulated in manuscript among scholars. Defence requirements, at a time when invasion was frequently expected, also demanded good plans of the countryside and cities of England, and some were made under orders from Lord Chancellor Burghley.

Many estate maps and town plans made at this period were realised in a combination of plan and elevation, in a kind of bird's eye view. It is thought that this style was derived partly from similar views in miniatures in illuminated manuscripts, partly from symbols such as the little buildings used earlier by Matthew Paris to denote towns, and partly from the molehills used to denote hills and mountains.

This type of presentation was used also on manuscript sea charts, numerous specimens of which had been made as a result of Henry VIII's concern in developing English sea-power. Such charts were frequently coloured and give charming bird's-eye views of English towns and villages, affording some indication of their appearance to contemporary eyes. As with so many other sea-charts, ships in full sail are frequently represented.

The views on these maps were each drawn on the spot from a ship anchored at given points along the coast. The draughtsman worked from the masthead, using an instrument known as a Kamal, cross-staff or Jacob's staff to measure the angles of the view. Kamal is an Arab word and indicates the middle-east origin of the instrument. It was first described in 1342 by a Provençal Jew, Levi ben Gerson. It consisted of a rod, some five or six feet long, along which a transom bar could slide. The rod was graduated in degrees.

One of the most important makers of charts such as these was Richard Popinjay. He was employed in the south coast and Channel Islands ports during the reign of Elizabeth I, from 1562 to 1587. His maps contain much detail, including dangerous rocks, quays, safe anchorages and roads inland. The maps are provided with scales.

The first engraved map of the British Isles which can be said to comply with post-mediaeval geographical standards, was produced in Rome in 1546. It has beneath its cartouche the initials G L A which some think are those of George Lily 'Anglorum' (d. 1559), an erudite Catholic exile then living in Italy, where he had become chaplain to Cardinal Pole. But it is probable that the map was produced under the influence of a number of Catholic exiles living in that country, for whom it was engraved by Italian craftsmen. Another theory is that the map was compiled in England and sent to Italy to be engraved. It is conceived on a conical projection, with longitude and latitude marked in the borders.

Though copper-engraving took a long time to reach this country, it was in due course introduced from the Low Countries by Thomas Gemini, a Belgian (*fl.* 1540–60). Gemini had access to the 'G L A' map and after altering it in several respects and adding his own name to it, he republished it in London. Only one impression of it remains, which is in the Bibliothèque Nationale at Paris. Gemini is mainly remembered for his authorship of a compendium of anatomy, for which he also engraved the plates. It is entitled *Compendiosa totius Anatomie delineatio*, and was first published in 1545 with a dedication to Henry VIII.

Flemish engravers played so large a part in sixteenth-century English cartography, that it would hardly be an exaggeration to say that it could not have existed without them. The first English sea atlas, a translation by Anthony Ashley of Lucas J. Waghenaer's *Spieghel der Zeervaert* (*Mariner's Mirrour*) was published in 1588, soon after the defeat of the Spanish Armada. It contained forty-five charts, with coastal profiles and with the usual complement of sea monsters and ships. Three of the charts were engraved by Hondius and nine by Theodore de Bry of Liège, both of whom were working in England at that time.

Renold or Renier Elstracke (1571–1625?), a Fleming, was one of the earliest and finest copper-engravers to work in England. Horace Walpole in his *Anecdotes of Painting in England* rather unkindly remarks that his 'works are more scarce than valuable'. Elstracke was a pupil of the English engraver William Rogers (b. *circa* 1545), who had himself been taught by Flemings: he was a pupil of the Wierix brothers of Antwerp, and engraved the title page to Jean Huygen Linschóten's collection of voyages to the East Indies. Elstracke engraved maps for Linschóten's *Itinerario*, and in 1619 he engraved William Baffin's map of the Mogul Empire, which had been compiled on information given by Sir Thomas Roe, the diplomat and explorer. But the most interesting of Elstracke's maps was a version, published in 1599, of the Italian Baptista Boazio's manuscript map of Ireland which had itself been completed by about 1588 (Plate 21). Boazio was a surveyor, probably brought back by the Earl of Essex from his expedition to Ireland in 1599. Boazio's map had been first published by Hondius in 1591, when it was engraved by Pieter van den Keere; the map in the 1602 and later editions of Ortelius was based on this version. It is an improvement on earlier maps of Ireland, in particular in its representation of the north and south-west areas. Only three copies of the 1599 edition exist, one of them printed on silk. Elstracke must have possessed more than his fair share of vanity, for he renamed one of the Aran Islands 'Elstracke's Isle'. He also signed the map in one of the lower corners and put the name of his publisher in the other: 'Mr Sudbury in Pope's Alley'. The map, which is on two plates, has five elaborate cartouches, some heraldry and much detail. It was dedicated to 'The most sacred and renowned my Gratious Soveraigne Elizabeth . . .', and the dedication explains how

> . . . Your Highness may distinctly see
> what Havens, Rockes, sands or Townes, in Ireland be.

The earliest map engraved by an Englishman is of the Holy Land. It was by Humfray Cole (1530?–91), who was a well-known maker of mathematical and scientific instruments. Some of these instruments have survived, including an astrolabe in the British Museum which once belonged to the elder brother of Charles I, the 'Incomparable' Henry, Prince of Wales. It is of splendid and ingenious workmanship and carefully and beautifully engraved.

Cole's map of the Holy Land was engraved for the second edition of Archbishop Matthew Parker's 'Bishop's Bible' of 1572, and a cartouche contains his name: 'Humfray Cole, goldsmith, a Englishman born in ye north and pertayning to ye Mint in the Tower, 1572.' Because he engraved this map, it has been claimed that Cole also engraved the portraits that appear in the same Bible, but there is no reliable evidence to support this.

The map contains the arms of Lord Burghley and has other decorations. One of these is a rebus of Richard Jugge, who printed it, a bird on a bush, with a scroll issuing from its mouth inscribed in italic: 'IUGGE, IUGGE, IUGGE, IUGGE'. There is also a scale showing both English and Italian miles, classical stadia and leagues. Symbols in the form of little buildings are used to denote towns, and heights are indicated by 'mole-hills'.

It is thought likely that Cole learnt the technique of engraving from Remigius Hogenburg, a Netherlandish engraver employed by Archbishop Parker. There is a further relationship with Hogenburg since Cole's map was copied from one, drawn by Tilmann Stolz and engraved by another member of the Hogenburg family, Franciscus.

As we have seen, the bird's-eye view was used by Popinjay and others to make it easy for pilots to recognise coastal features. It was also used on inland maps and town plans, doubtless with much the same idea in mind, although the intention was certainly in part artistic. The first town plan of this type was published in 1574 under the patronage of Archbishop Parker. It was the work of Richard Lyne, a pupil of Franciscus Hogenburg, and is a plan of Cambridge, set within a decorative border marked with the four cardinal points of the compass in cartouches. It appeared in Dr. John Caius's *History of the University* (1574), and with the town shown in plan as well as in elevation we can thus gain an excellent idea of what Cambridge looked like in those days, and a delightful, almost arcadian scene it must have presented (Plate 22). In many ways the plan is reminiscent of certain Indian and Persian miniatures which also show their scenes at once in plan and elevation. It has, too, their air of poetry, which is accentuated by the town's various knot, physic and topiary gardens, the grazing animals in the surrounding meadows, and the many little streams that may still be seen in Cambridge to this day, making it, in a limited way, a water city like Bruges. Two large cartouches contain a description of Cambridge and a key to the buildings depicted. There are also coats of arms, swags of fruit, masks, flowers and dragonflies.

Apart from this plan, Lyne also engraved in 1574 a large genealogical chart of the history of Great Britain. This was partly engraved by Hogenberg. Lyne was also a well-known painter in his day, and it is thought that he may have painted a portrait of Archbishop Parker that was later engraved by Hogenberg.

The sixteenth century saw many advances in cartographical science in England as well as on the continent. One of those responsible for such advances was William

Cuningham or Keningham (b. 1531), engraver, astrologer and physician. He studied at Cambridge and Heidelberg; he was probably made an M.D. at the latter university. At about the same time he changed his name from Keningham to Cuningham. He returned to England and from 1556 to 1559 lived at Norwich. After this he lived in London, at Coleman Street, where he enjoyed a wide reputation in both medicine and astrology. In 1563 he became a lecturer at Surgeons' Hall.

Cuningham wrote many books, including one or two *Newe Almanackes and Prognostications, The Astronomical Ring, Organographia, Gazophilacion Astronomicum, Abacus, or a Book of Longitudes and Latitudes of various places,* and, most important of all to the subject of this book, *The Cosmographicall Glasse, conteyning the pleasant Principles of Cosmographie, Geographie, Hydrographie, or Navigation* (1559). Among its plates is a map of the city of Norwich (Plate 20).

One of Cuningham's most important contributions to cartography was the introduction of continental methods which he had doubtless seen at Heidelberg. Though he was primarily a physician he pioneered several new techniques in cartography and surveying. The equipment he used included a surveying chain, an astrolabe and a compass.

Another important figure of the time was Leonard Digges (d. *circa* 1571), a mathematician and a member of an ancient family which had produced members of parliament, sheriffs, magistrates and judges. He studied at University College, Oxford, but left without taking a degree. But he spent a great deal of his time on scientific pursuits, especially land-surveying and mathematics, and his lack of academic distinction did not prevent him from attaining later success in these subjects. He also became a fine architect, being described by Thomas Fuller, the seventeenth-century author and divine, as 'the best architect in that age, for all manner of buildings, for conveniency, pleasure, state, strength, being excellent at fortifications'.

Like Cuningham, Digges was a prolific writer and produced several books, including the most famous of them all, *A Geometricall Practise, named Pantometria, divided into Three Bookes, Longimetria, Planimetria, and Stereometria, containing Rules manifolde for Mensuration of all Lines, Superficies and Solides.* This was completed and published by Digges's son in 1571. It contains a description of a primitive theodolite (an instrument used in surveying for measuring angles), and a large section on optics, which is of great importance in accurate surveying. He refers in this section to the use of a series of lenses for magnifying, and he is thus held to have anticipated the invention of the telescope.

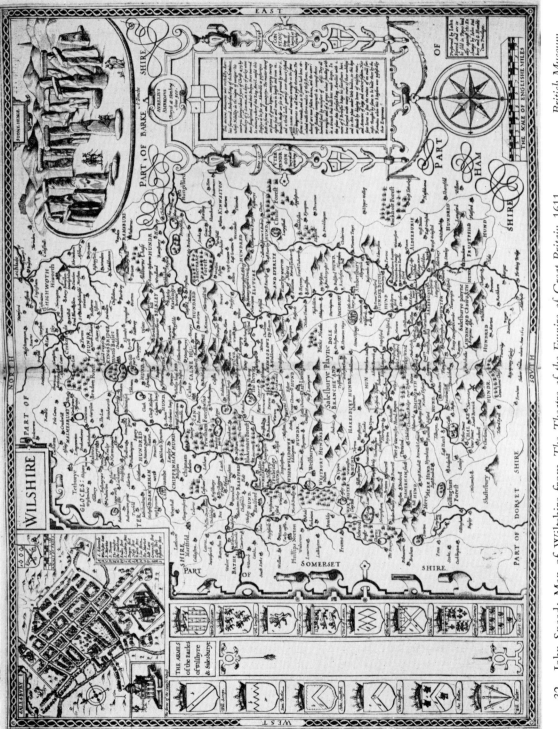

32. John Speed: Map of Wiltshire from *The Theatre of the Empire of Great Britain*, 1611.　　*British Museum.*

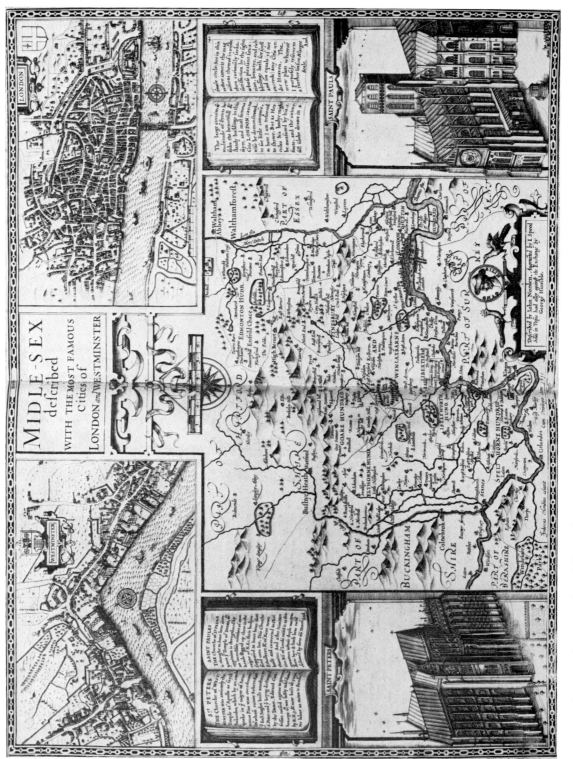

33. John Speed: Map of Middlesex from *The Theatre of the Empire of Great Britain*, 1611.

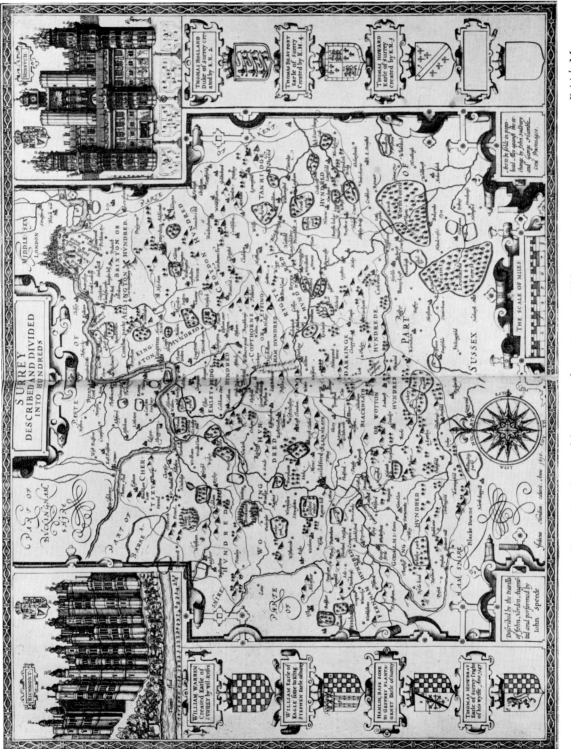

34. John Speed: Map of Surrey from *The Theatre of the Empire of Great Britain*, 1611.

British Museum.

and Drayning of Divers Fenns and Marshes, 1662.   University Library, Cambridge.

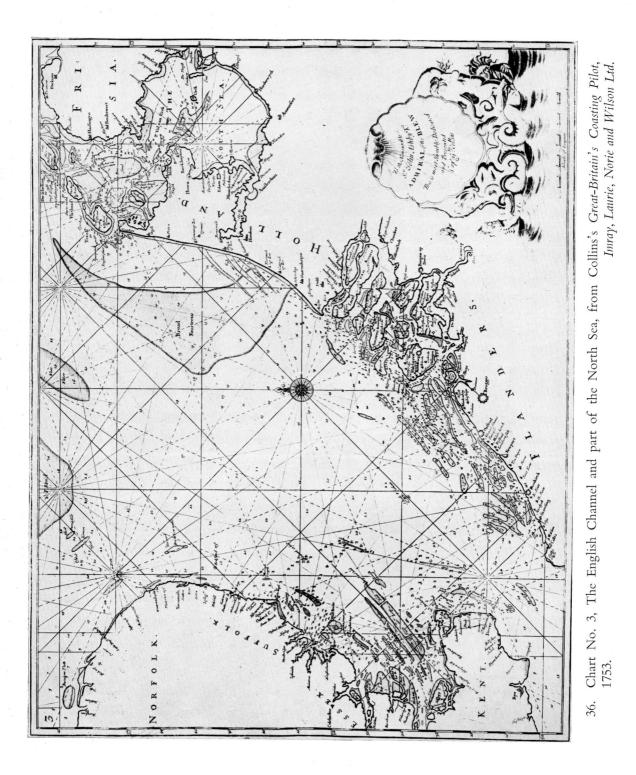

36. Chart No. 3, The English Channel and part of the North Sea, from Collins's *Great-Britain's Coasting Pilot*, 1753.
Inray, Laurie, Norie and Wilson Ltd.

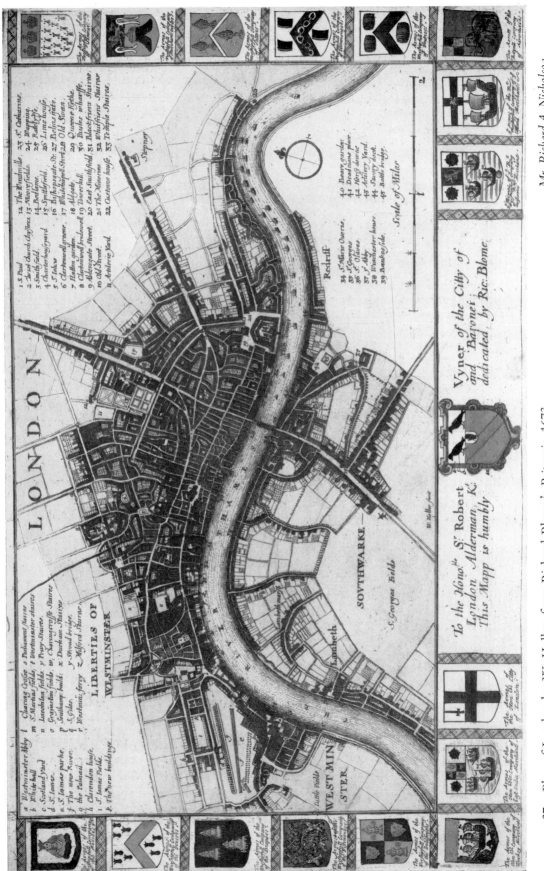

37. Plan of London by W. Hollar, from Richard Blome's *Britannia*, 1673.

Mr. Richard A. Nicholso 1.

38. The Road from London to Slough, from *Britannia Depicta or Ogilby Improved*, 1720. Engraved by Emanuel Bowen.   *Private Collection.*

At this period land itineraries first began to be published in England. These were terrestrial portolans, so to speak, giving tables of various routes, including the towns through which one might pass on certain journeys, and the mileages between each— a kind of cartographic abstract.

These, then, were some of the details of the background against which the great English cartographers of the sixteenth century began to practise their work. One of the greatest of these early cartographers, and a man who produced some of the most attractive maps of the time, was Christopher Saxton (*circa* 1542–*circa* 1610).

Christopher Saxton, the 'Father of English cartography', was born at Dunningley, near Dewsbury in the West Riding of Yorkshire; he came of an old local family. He was probably educated at Cambridge, but this is uncertain. He came to London at an unknown date and became attached to the household of Thomas Seckford (1515?– 88), lawyer, Master of Requests and of the Court of Wards, and M.P. for Ipswich. This was a propitious association, for it was at the request and expense of Seckford that Saxton undertook to survey and draw maps of the English and Welsh counties. The maps were begun in 1574 and completed in 1579; they were published in that year and dedicated to Queen Elizabeth, who had given her authority to the undertaking. Consequently the royal arms, in addition to those of Seckford, appear on each map (Plates 23, 24 & 26).

A ten-year privilege to make and market the maps was granted to Saxton in 1577. He was granted a coat of arms by the Queen—the only English cartographer ever to have received such an honour from the reigning monarch—for his 'Geographicall discription of all the several Shires and counties within this Realme now finished to his own lasting praise'.[1] In 1574 the Queen also granted Saxton a lease of certain lands at Grigston Manor in Suffolk 'in consideration of his expenses lately sustained in the survey of divers parts of England', and in 1575 he was given the reversion of the receivership of lands that had once belonged to the London Hospital of St. John of Jerusalem. Later still, in 1580, the Queen assigned to Saxton lands in the parish of St. Sepulchre without Newgate in London.

Seckford's part in the undertaking was commemorated by a copper penny token issued in 1796 by R. Loder of Woodbridge in Suffolk. It bears a portrait of Seckford, and, on the reverse, his arms and a legend, part of which reads 'THO: SEKFORD ESQ FOUNDED WOODBRIDGE ALMSHOUSES 1587. AT WHOSE EXPENCE COUNTY MAPS WERE FIRST ENGRAVED 1574' (Plate 28). Seckford belonged to an old Suffolk family, whose

---

[1] Saxton's arms were three chaplets or garlands in a bend gules.

seat was Seckford Hall, near Woodbridge. He was a man of wide interests, and in 1587, the year of his death, founded almshouses for thirteen poor unmarried men and their nurses, three poor women. The inmates wore a badge, the design of which was based on Seckford's arms. The almshouses were financed by income from Seckford's estate in Clerkenwell, and some think that this income was also used to finance Saxton's survey.

Despite Seckford's patronage, difficulties attended Saxton in his survey. To help overcome some of them he was given an order from the Privy Council that he was 'to be assisted in all places where he shall come for the view of such places to describe certein counties in cartes, being thereunto appointed by her Majestie's bill under her signet'. Wales presented particular problems, for in those days it was a wild and remote country where English was little spoken, and a special order was issued in 1576 to all mayors, justices of the peace and other authorities, to ensure Saxton was 'conducted unto any towre, castle, highe place or hill, to view that countrey, and that he may be accompanied with ij or iij honest men, such as do best know the countrey, for the better accomplishment of that service; and that at his departure from any towne or place that he hath taken the view of, the said towne do set forth a horseman that can speke both Welshe and Englishe, to safe-conduct him to the next market-towne'.

The mention of 'towre, castle, highe place or hill' is a reminder that in those days surveys were usually begun from such elevations, using triangulation with the assistance of cross-staff and compass. The order also tells us something of Saxton himself: that he was a field worker who obtained his own data and was not merely satisfied (as would have been all too easy, indeed customary, at that time) to copy what somebody else had drawn or written.

Nevertheless, although Saxton was above all original, he could not have assembled the whole of the material for his maps without drawing on other topographical authorities, such as Leland's itinerary and William Lambarde's *Perambulation of Kent*. In turn, Lambarde may have had help from Saxton in compiling a map of Kent which he made in 1570. Saxton's partial reliance on existing authorities is especially evident when one considers the speed at which he worked; his survey of Wales plus the printing of the maps were completed in just over two years, an incredibly short time if one assumes he was working more or less single-handed. To complete such a survey and print the maps in two years would be an outstanding achievement even today with modern instruments and methods of production, and with a number of teams collaborating to take the measurements. Yet Saxton was, according to sixteenth-century standards, outstandingly accurate. It is thought that Seckford may have supervised the engraving and printing of the maps while Saxton was in the country surveying.

Saxton did not show roads on his maps. They did not appear on English maps until John Norden compiled his map of Middlesex in 1593 (pages 62–3).

The earliest maps in Saxton's atlas were those of Berkshire, Buckinghamshire and Oxfordshire (which appeared on one map) and Norfolk. The former is not signed by the engraver, but the latter was signed and engraved by Cornelis de Hooge or Hogius. They were made in 1574. Six further maps depicting nine counties bear the date of 1575. These were engraved by Remigius Hogenberg and Leonard Terwoort. Terwoort also engraved the map of Suffolk, but he did not sign it, and there is certain stylistic evidence that in addition he engraved the maps of Berkshire, Buckingham-shire, and Oxfordshire and Northamptonshire. Eight maps showing fifteen counties are dated 1576 and on two of these—Durham and Westmorland—is found the name of the first great English map engraver, Augustine Ryther. He sometimes signed himself as 'Augustinus Ryther Anglus', to show that he was English and not one of the more commonplace Netherlandish engravers. The map of Westmorland, too, saw Seck-ford's motto changed from *Pestis patriae pigrices*, as had appeared on the earlier maps, to *Industria naturam ornat*. All subsequent maps bear the new motto.

In the following year, 1577, the remaining maps of the English counties, which included the big double-sheet map of Yorkshire, were completed as well as two of the Welsh counties, Flint and Denbigh. Among the engravers of the 1577 maps were two more English engravers, Francis Scatter (Staffordshire, Cheshire, and the unsigned map of Radnorshire, Brecknockshire, Cardiganshire and Carmarthenshire—the latter made in 1578) and Nicholas Reynolds (Hertfordshire). In 1578 came five sheets depicting ten Welsh counties. The general map of England and Wales, and the frontispiece and index were added in 1579.

The frontispiece is one of the most magnificent to be found in any atlas. It is not known for certain who engraved it, but some consider it to be the work of Ryther; others attribute it to Hogenberg. The Queen is shown enthroned as patroness of Astronomy and Geography, which are represented on each side of her classical pavilion by bearded figures holding terrestrial and armillary spheres. The Queen herself, almost a hieratic figure, is seated with orb and sceptre, and is represented as she appears in some contemporary miniatures by Nicholas Hilliard, the centre and sun of her age. Above the pavilion are the royal arms of England, flanked by putti grasping wreaths of laurel and roses. A lower tablet contains an elaborate cartouche, which contains a colophon and is decorated with sprays of fruit and groups of trophies.   This is flanked by two further figures representing Astronomy and Geography.   The whole is contained within a decorative border (Plate 23).

The maps themselves are also enclosed in decorative borders.   The title cartouches are richly decorated, and the lettering itself is a splendid italic.   There are the customary ships and sea monsters around the coasts.   The conventional 'molehills' are used for portraying relief, with here and there some strange inconsistencies—Snowdon is shown three miles high (Plate 24).   Towns and cities are represented by little groups of buildings, graded according to size, wooded areas by little 'Noah's ark' trees, parks by palings and trees, and important rivers by double lines interrupted at bridges.   It is interesting to note that of eight hundred and seventeen parks shown by Saxton, many still exist.[1]

The index to the atlas was issued in three separate editions.   They may easily be identified by the fact that in the first the names of the counties are arranged in a column, with the numbers before them; in the second the title is narrower, and the list is contained in a line framework; the index in the third edition is longer and more elaborate than the others, and has coats of arms and tables of towns, and other details.

Although it is a valuable book, Saxton's atlas is not really rare for a sixteenth-century publication.   There are many variants among surviving copies, which fact seems to point to frequent issues.   Single copies of the maps were also sold.

There are some major inconsistencies in Saxton's work.   The scales, for instance, vary considerably from map to map.   To take two examples, Hertfordshire is drawn one and three-quarter miles to one inch, Cornwall just under four miles to one inch. It is clear that Saxton must have found some difficulty in dealing with the contemporary practice of using 'customary miles', which varied from place to place as much as four furlongs.   Saxton varied his scales accordingly.   The old English mile was reckoned, according to the place, at about 2,140 yards, as against the present-day mile of 1,760 yards.

Saxton's *Atlas* was issued, with various alterations, until about 1749.   The earlier editions showed slight differences, such as in watermarks and in the coats of arms represented, but in many cases are impossible to date accurately.   An edition was published in 1645 by the obscure London publisher, William Web.   In this the maps had the date 1642 inserted, and the monogram CR substituted for the earlier ER on the maps of Cornwall, Durham and Gloucester and on that of England and Wales.

[1] See Hugh Prince's book in the Bibliography (page 86).

Hundreds[1] had been marked on only five of the original Saxton maps: Cornwall, Essex, Hertfordshire, Norfolk and Suffolk. Web added them to the maps of Lincolnshire and Nottinghamshire and inserted their boundaries on the Yorkshire map. Plans of the cities of Hull and York were also added to the Yorkshire map (the one of York copied from a map by John Speed), and a plan of Berwick, also copied from one by Speed, was added to the map of Northumberland.

More of these plans from Speed's atlas were added about 1689 to twenty-four further Saxton maps. These were sold by the London publisher, Philip Lea (*fl.* 1666–1700). Lea also completed the hundreds on the Yorkshire map by adding the names, and he also added hundreds to the remainder of the maps, but not all at the same time. In other cases arms were erased and—in the case of the Yorkshire map—the town plans. From 1689 the maps were issued as an atlas by Lea under the title *All the Shires of England and Wales Described by Christopher Saxton. Being the Best and Original Mapps. With many Additions and Corrections by Philip Lea.* A variant of this has the word 'All' omitted from the beginning of the title, an 'Explanation' of the symbols used, and other minor differences. The maps in this atlas show roads, although these are not always present on the separately issued maps. The plates subsequently came into the possession of George Willdey, mapseller and toymaker, who in 1725 reprinted thirteen of the maps. There are many more minor differences in addition to those mentioned, but they will be enough to indicate the extreme difficulties and some of the pitfalls that await the collector.

In addition to the various reprints from the original plates, reduced versions of Saxton's maps are known, such as those that were reprinted in Camden's *Britannia* in 1607, 1610 and 1637.

A rare work by Saxton is his large-scale map of England and Wales in twenty-one sheets, engraved to a scale of eight miles to one inch. Only two copies are known, one in the British Museum, the other in Birmingham Public Library. Lea published a reproduction of it, with alterations, in 1687, but this, too, is rare. Further reprints were made by various publishers until as late as 1795, but alterations and additions were made to most of them.

John Norden (1548–1626), topographer and surveyor, is chronologically the next important figure in English cartography. His father was a gentleman and lived in Middlesex, but John was born probably in Somerset or Wiltshire. In later years he lived near Fulham, where he had a garden; his friend John Gerard, the famous herbalist,

---

[1] A hundred is a division of a county, originally supposed to have contained one hundred families.

presented him with some red-beet seeds 'altogether of one colour', and they 'in his garden brought foorth many other beautiful colours'.

Norden is first heard of in a Privy Council order from Hampton Court, dated 27 January 1593, 'To all Lieuts, etc., of Counties'. It declared that the bearer, 'John Norden, gent.,' was 'authorised and appointed by her Majesty to travil through England and Wales to make more perfect descriptions, charts and maps'. This resulted in Norden's first work, *Speculum Britanniae, firste parte . . . Middlesex*, which was published in the same year (Plate 31). The second part, Hertfordshire, followed in 1598. Another order, issued by Lord Burghley in July 1594, spoke of 'The bearer, John Norden, who has already imprinted certain shires to his great commendation, and . . . intends to proceed with the rest as time and ability permit'. The remainder of his work, despite his great industriousness, was destined to remain unknown until he had been dead for three hundred years. Norden was the first Englishman to plan an exhaustive series of county histories, and there is in the British Museum a manuscript by him entitled *A Chorographical Discription of the severall Shires and Islands, of Middlesex, Essex, Surrey, Sussex, Hamshire, Weighte, Garnesay and Jarsay, performed by the traveyle and view of John Norden, 1595.*

Norden's work was bedevilled by financial difficulties. This may be gathered from the dedication, to Burghley, of his book *Preparative to . . . Speculum Britanniae* (1596), which he wrote from his 'poore house neere Fulham', and in which he mentions his 'struggle with want'. He had spent nearly all he had in producing the two parts of his *Speculum*. In 1597 Norden made two manuscript copies of his *Description of Hertfordshire*, one of which he dedicated to Burghley and the other to Queen Elizabeth. To the Queen he wrote: 'I was drawne unto them by honorable Counsellors and warranted by your royall fauor, I was promised sufficient allowance and in hope thereof onlie I proceeded . . . By attendaunce and . . . travaile in the business I have spent aboue a thousand markes and five years time . . . Onlie your Maiestie's princelie fauor is my hope without which I myself most miserablie perish, my familie in penurie and the work unperformed.' Apparently both the Queen and Burghley hardened their hearts, for Norden does not appear to have received a reply. He therefore published the Hertfordshire book at his own expense. It contained a map engraved by William Kip.

Apart from the surveys of Hertfordshire and Middlesex mentioned above, Norden completed in manuscript surveys of five other counties: Essex, Northamptonshire, Cornwall, Kent and Surrey. Only Hertfordshire and Middlesex were published during his lifetime. His survey of Cornwall was published in 1728, that of Essex in 1840. The *Description of Cornwall* contains over a hundred pages and has 'a Map of the County and each Hundred; in which are contained the Names and Seats of the several Gentlemen then Inhabitants: as also, thirteen Views of the most remarkable Curiosities in the County'. It is divided into two-inch squares by a series of vertical and horizontal lines, the horizontal lines numbered and the vertical spaces lettered (Plate 25). In this way the position of St. Michael's Mount, for example, may be identified by stating that it is in the eastern part of the square 10, 12 d. The grid also provides a supplement to the scale, for the squares are of constant size. Such grids had been used in European cartography since about 1525.

By 1600 Norden had become Surveyor of the Crown Woods and Forests in Berkshire, Devonshire and Surrey and, in 1605, Surveyor of the Duchy of Cornwall. He was a prolific writer and in addition to the works already mentioned, compiled a set of distance tables which were published in 1625. They were entitled *England, an intended Guyde for English Travilers*, and were intended to be used in conjunction with Speed's county maps. In addition to the table, each plate contained a thumb-nail map of the district and a scale with a pair of dividers open on it.

In the distance tables there is a triangular 'grid' containing squares with numbers. Along each of two edges of the triangle are names of cities and towns. To find the distance, say, between Abingdon and Wallingford on the Berkshire table, the name of Abingdon is found on the top edge, the name of Wallingford on the outer edge, and the square where the two rows meet will give the distance in miles, in this case seven (Plate 27). The idea is still used today.

These maps were copied several times by Jacob van Langren between 1635 and 1643 and published under the title *A Direction for the English Traviller*. They were used also, between 1657 and 1677, by J. Jenner in his *Book of Names*, and in 1720 by Thomas Cox in his *Magna Britannia*.

Other works by Norden include *Surveyor's Dialogue* (1607; republished 1610, 1618, 1758) and numerous works in manuscript. He was a practising craftsman and in addition to illustrating his own works with maps and plans, made maps for Camden's *Britannia* (Hampshire, Hertfordshire, Kent, Middlesex, Surrey and Sussex) and for John Speed (Cornwall, Essex, Middlesex, Surrey and Sussex). He also made some important town plans, such as his bird's-eye plans of Chichester, Higham Ferrers, London and Westminster, some of which were engraved by Pieter van den Keere. There was in addition the now lost view of London, of which John Leland wrote in his

De Rebus Brit. Collecteana (first published at Oxford in 1715): 'Mr. Norden designed a "View of London" in eight sheets, which was also engraved. At the bottom of which was the Representation of the Lord Mayor's Show, all on Horseback. . . . The View was taken by Norden from the Pitch of the Hill towards Dulwich College going to Camberwell from London, in which College, on the Stair Case, I had a sight of it. Mr. Secretary Pepys went afterwards to view it by my recommendation, and was very desirous to have purchased it. But since it was decayed and quite destroyed by means of the moistness of the Walls. This was made about the Year 1604 or 1606 to best of my memory, and I have not met with any other of the like kind.'

Norden's maps are attractive specimens of the craft (Plate 25). They have the usual complement of ships and monsters, and sometimes contain inset bird's-eye view plans of towns. His treatment of conventional symbols such as trees and hills, is similar to Saxton's. The lettering is clear and pleasing. One innovation that he made was the inclusion of roads. He also gave details of industry, agriculture and battles. For the latter he often used the conventional symbol of crossed swords, but on his map of Hertfordshire, figures of fighting soldiers were used. And he added a special symbol for 'decayde places': villages and other places, that is, that had vanished.

Before passing on we must note one or two figures of some importance whose output, however, was very limited. Among them was the herald William Smith (1550–1618), Rouge Dragon Pursuivant. He was author of *The Vale Royall of England, or County Palatine of Chester; containing a Geographicall Description of the said Countrey or Shyre* . . . (1585), and *The Particular Description of England, with Portraitures of certaine of the chiefest Cities and Townes* (published in 1879, over two centuries after his death). Apart from this he left a large amount of manuscript material, much of it topographical. The manuscript of *The Particular Description of England* contained a map which was reproduced in the 1879 printing. Apart from this, Smith published a map of Cheshire (1585), based on Saxton's but corrected. It contained a folding view of Chester and a view of Halton. In addition Smith published a map of Lancashire (1598), which was also based on Saxton's. Another herald cartographer was Robert Glover (1544–88), Portcullis Pursuivant from 1567 and Somerset Herald from 1571. He made a good manuscript of Kent.

A map of Kent on two large sheets was made in 1596 by Philip Symonson. Of this first issue only one sheet, that of the eastern half of the county, has survived, but there were subsequent editions. It was engraved by Charles Whitwell (*fl.* 1593–1606), and like Norden's maps, showed main roads.

Yet another herald with close connexions with cartography was William Camden (1551–1623), antiquary, historian and Clarenceux King of Arms. He was educated at Christ's Hospital, St. Paul's School and Oxford, where he founded a chair. Camden was author of the great antiquarian investigation of Britain, *Britannia*, first published in 1586. Several further editions followed, the last Latin edition to be published in Camden's lifetime being that of 1607; it was also the first to contain a group (fifty-seven) of county maps, a regular feature in subsequent editions. The maps, which have engraved surfaces of approximately 14 by 11 inches, are based on those of Saxton and Norden; they were engraved by William Kip and William Hole.

The 1607 edition has a Latin text on the reverse of the maps. The 1610 edition was the first to appear with an English text, which was separately engraved, the reverse sides of the maps being left blank. Other editions include the following: 1637 (fifty-seven maps), 1695 (fifty maps, engraved by Sutton Nicholls and John Sturt), 1715, 1722 (two volumes), 1735, 1772 and 1789 (three volumes; sixty maps engraved by J. Cary). An abridged edition was published in 1617; it contained reductions of Saxton's maps by Peter Keere. A further abridged edition was issued in 1626 with fifty-one small maps; this was printed by John Bill. The maps of the 1789 edition were used in 1805 for John Stockdale's *New British Atlas*. *Camden's Britannia epitomised and continued* was published by Bohn in six volumes in 1845.

Eighteen unusual maps engraved by William Hole appeared in *Poly-Olbion, or a Chorographicall Description of the Tracts, Rivers, Mountaines, Forests, and other Parts of Great Britain* (1613), by Michael Drayton (1563–1631). They tell us little about the countryside they are supposed to portray, but are poetic interpretations in which nymphs, shepherds, naiads and demigods people the shires of England, in which they represent towns, spas, hills and other features. Windsor forest is depicted by a group of trees with a draped female figure beside it holding a bow and an arrow. A horse prances over Berkshire's Vale of the White Horse, far removed in form from the primitive beast that is actually there (Plate 30). Rivers are crowned naiads. On each bank of the widening Severn an orchestra of nymphs plays on lutes, viols, harps, bagpipes and a portative organ; one nymph holds aloft a banner with the legend 'St. George for England', an allusion no doubt to the close proximity of St. George's Channel. Farther out in the Channel a portly naked Neptune wielding a trident, stands on a sea-monster (Plate 29). It is not geographical England, but the England of *The Faerie Queene* and, as it should be, of Drayton's *Poly-Olbion*:

E

Beare bravely up my Muse, the way thou went'st before,
And crosse the kingly Thames to the Essexian shore,
Stem up his tyde-full streame, upon that side to rise,
When Cauney, Albion's child in-iled richly lyes,
Which, though her lower scite doth make her seem but meane,
Of him as dearly lov'd as Shepey is or Greane,
And him as dearly lov'd: for when he would depart,
With Hercules to fight, she tooke it so to heart,
That falling low and flat, her blubbered face to hide,
By Thames shee welneere is surrounded every tyde:
And since of worldly state, she never taketh keepe,
But onely gives her selfe, to tende, and milke her sheepe.

One of the most famous of all English cartographers was John Speed (1552–1629), the son of a merchant tailor. He was born in Cheshire and brought up in his father's trade, which was carried on in London. In 1580 he was admitted to the freedom of the Worshipful Company of Merchant Taylors. Two years later he married and settled in Moorfields, London, where he rented premises and a garden from his livery company at twenty shillings a year. On this he built a 'fayer house which may stand him in 400 *l*', and took over further adjacent land from the Company at two pounds a year. In 1614 he took over still more land in a lease of a prebendal estate held by the Merchant Taylors' Company from St. Paul's Cathedral. This lease was obtained by Speed largely through Sir Fulk Greville, first Lord Brooke. Speed refers to Greville's generosity in his atlas, *Theatre of the Empire of Great Britaine*, saying: 'whose merits to me-ward I do acknowledge, in setting this hand free from the daily imployments of a manuall trade, and giving it full liberty thus to express the inclination of my mind, himself being the procurer of my present estate'. Greville put forward further recommendations on Speed's behalf in 1598, when he obtained for him 'a waiter's room in the custom-house'.

Speed's cartographic activities were begun in his spare time. He made and presented maps to the Queen in 1598 and to the Merchant Taylors' Company in 1600. Of these acknowledgement was made of his 'very rare and ingenious capacitie in drawing and setting forthe the mappes and genealogies, and other very excellent inventions'. Of his published maps one, a copy of Norden's map of Surrey, was made in 1607 for Camden's *Britannia*, and between the following year and 1610 he published over fifty separate maps of England and Wales, some after Norden, the others after Saxton. In 1611 these appeared in the first edition of the *Theatre of the Empire of Great Britaine*, published by George Humble, who had three years previously received a licence for them. There were many subsequent editions, which we shall shortly consider.

Despite his work on maps, Speed was not really a cartographer, but a historian who compiled maps from other men's data. He said of himself, 'I have put my sickle into other men's corne', for he made no real contributions to the development of the craft. Yet he always acknowledged his sources, which cannot be said of many cartographers; and he constantly revised his work in view of fresh information.

Speed's maps are full of historical and antiquarian detail, and above all are decorative, with their inset views and plans, allegorical personages, cartouches, heraldry and decorative borders. The colouring, when it is contemporary, is magnificent, but the maps were usually published uncoloured; and later, even modern, colouring is common. The clear and decorative lettering is a mixture of roman and italic, and on some of the larger inscriptions swash letters occur. The maps were engraved by Jodocus Hondius the elder.

Many of the views on the maps are splendid. Especially so is Stonehenge ('Wilshire') with its group of trilithons, much as it appears today, except that here are but five viewers in Jacobean dress, and one dog, where today there are often crowds (Plate 32). More formal is the view of Buxton ('Darby-shire') with its 'faire square building of free-stone' and its well 'called *S. Anne's*; neere unto which another very cold spring bubled up'. Old St. Paul's is shown on the map of Middlesex (Plate 33). Most splendid of all is the view of the palace of Nonsuch (Surrey), with its parterred garden, drawn tipped up as in a Persian miniature or an English sampler (Plate 34). Battle scenes are common, with infantry drawn up into square formations, in full battle array, with banners and lances. In other places appear occasional representations of antiquities and coins.

The town plans were taken mainly from the work of other cartographers, including William Cuningham, William Mathew, John Norden, Christopher Schwytzer and William Smith. Speed claimed as his own the plans of St. Davids and Pembroke.

Speed also issued maps of foreign parts and his *Prospect of the most famous parts of the world* (1627 and subsequent editions) was the first general atlas to be published by an Englishman. The style of its maps follows closely that of those in the *Theatre*.

Speed's descriptive passages, which in some editions are printed on the back of the maps, are splendid, and often poetic. That for Herefordshire gives us details of a curious event that occurred in Speed's lifetime:

But more admirable was the worke of the Omnipotent even in our owne remembrances, and yeere of Christ Iesus 1571.   When the *Marcley Hill* in the East of this Shire rouzed it selfe out of a dead sleepe, with a roaring noise removed from the place where it stood, and for three days together travelled from her first site, to the great amazement and feare of the beholders. It began to iourney upon the seventh day of February, being Saturday, at six of the clocke at night, and by seven in the next morning had gone fortie paces, carrying with it Sheepe in their coats, hedge-rowes, and trees; whereof some were overturned, and some that stood upon the Plaine, are firmely growing upon the Hill, those that were East, were turned West; and those in the West were set in the East: in which remove it overthrew *Kinnaston Chappell*, and turned two high-waies neere a hundred yards from their usuall paths formerly trod.   The ground thus travelling, was about twenty six Acres, which opening it selfe with Rockes and all, bare the Earth before it for foure hundred yards space without any stay, leaving that which was Pasturage in place of the Tillage, and the Tillage overspread with Pasturage.   Lastly, over-whelming her lower parts, mounted to an hill of twelve fadomes high, and there rested herselfe after three daies travell: remaining His marke that so laid his hand upon this Rock, whose power hath poised the Hils in his Ballance.

Speed's maps were widely used by the armies in the Civil War.   For this purpose they were sewn together in a vellum cover and folded so as to go in a coat pocket. They were said to be 'Usefull for all Comanders for Quarteringe of Souldiers & all sorts of Persons, that would be informed, Where the Armies be'.

Speed was elected a member of the Society of Antiquaries.   He met Camden and other members of the Society, who helped and encouraged him to write *The History of Great Britaine under the conquests of ye Romans, Saxons, Danes and Normans* . . . which was dedicated to James I and published in 1611.   Several other editions followed.   The work established Speed's reputation as an historian.   Other books by him were *Genealogies recorded in Sacred Scripture* (*circa* 1611), of which thirty-three editions appeared before 1640; and *A Cloud of witnesses* . . . *confirming unto us the Truth of the Histories in God's most Holie Word* (1616).

Towards the end of his life Speed became blind and suffered from gallstones, which did not, however, deter him from continuing to write.   He died on 28 July 1629 and was buried in St. Giles's Cripplegate.

☆   ☆   ☆

Here are brief details of various editions of Speed's *Theatre*:

The Theatre of the empire of Great Britaine: *presenting an exact geography of the kingdomes of England, Scotland, Ireland, and the isles adioyning: with the shires, hundreds, cities and shire-townes, within ye kingdome of England, divided and described by John Speed.   Imprinted*

*at London. Anno, cum privilegio, 1611 [-1612]. And are to be solde by John Sudbury &
George Humble, in Popes-head alley at ye signe of ye white horse.* fol. 9½″ × 15″. In four
parts, comprising books 1 to 4 of Speed's *History of England.* 67 maps. Isle of Man
by Thomas Durham; Isle of Wight by William White. Remainder by John Speed
from surveys of Norden and Saxton. Eng. mostly by Jodocus Hondius. They have
English text printed on their backs. It is a shortened version of Camden's *Britannia,*
except in the case of Norfolk; this is by Sir William Spelman. Printed by W. Hall.
Most of the maps bear the imprint of Sudbury and Humble, some of Humble alone.

Before the issue of the above atlas the maps had, between 1605 and 1610, been
issued separately, without text on their backs. A collection of these is in the British
Museum, and in many cases they were printed before the engraver had added his name
and dates. But note that the 1713, 1743 and 1770 edns. are also without text on their
backs.

Some other editions with brief details of some of their variations:

1614 (Sudbury and Humble). Date on title page altered to 1614. With a sheet
following title page, listing the parts of the 'empire', and 'The Achievement of our
Soveraigne King James . . .' This is followed by a dedication to the King, the
Royal Arms, four leaves addressed to the reader, verses of J. Sanderson and index
to the maps.

1616 (Sudbury and Humble). *Theatrum Imperii Magnae Britanniae* . . . The only edn.
with Latin text. Trs. by Philemon Holland. This edn. is of great rarity.

1627–31 (George Humble). Includes, facing the title page, a portrait of Speed, eng.
by S. Saevry. Text, incl. preliminary pages, reset in different type, with double-
line border.

1650–62 (Roger Rea the elder and younger). The whole of the text reset in smaller
type. New ornamental band at top of 'The British Empire' and following pages.
Circular space instead of shield in middle of band. In the preliminary pages the
author's address and the verses each occupy a whole sheet. In part I imprint on
maps altered from 'J. Sudbury and G. Humble' to 'Roger Rea'.

1676 (Thomas Bassett and Richard Chiswell). Title page re-engraved, with many
alterations, by R. White. Arms of Charles II replace those of James I in preliminary pages. Text completely reset. New initials on the text at back of maps.

Imprints now thus: 'Sold by Thomas Bassett in Fleetstreet, and by Richard Chiswell in St. Pauls Church yard.'

1713 and 1743 (Henry Overton).   Title page is printed from the eng. plate of 1611 edn., but original title replaced by this: *England fully described in a compleat sett of mapps of ye county's of England and Wales, with their islands, containing, in all, 58 mapps.*   In the 1743 edn. the words 'by John Speed' follow '58 mapps'.   In both edns. maps are reprints of the 1676 edn. above, but with main roads eng. on all plates.   Also the Bassett and Chiswell imprint is replaced by this: 'Henry Overton at the white horse without Newgate, London.'   Maps have plain backs.

1770 (C. Dicey and Co.). $11\frac{1}{2}'' \times 17\frac{1}{4}''$.   Title now *The English atlas, or a complete set of maps of the counties in England and Wales.*   Of the maps, 47 are from the Overton edn. of 1743 (*see above*); 5 are copied from maps by Jansson, 4 of which have the imprint 'Sutton Nicholls sculp.'; 2 others are by the Overtons, 1 by J. the other by H., and there are 2 road maps.   All but 4 have the imprint of C. Dicey and Co. Maps have plain backs.

Complete details of the foregoing editions of Speed's *Theatre* will be found in Chubb, *The Printed maps in the atlases of Great Britain* (see *Bibliography*).[1]

☆   ☆   ☆

Some interesting seventeenth-century maps were made by the Lancashire antiquary and topographer Sir William Dugdale (1605–86), Garter King of Arms.   They include maps of the Fens (1662, 1772; engraved by W. Hollar; Plate 35), Romney Marsh (1662, 1772) and Warwickshire (1656, 1765; engraved by R. Vaughan).   The Fens map appeared with others in his great work *The History of Imbanking and Drayning of Divers Fenns and Marshes*, and that of Warwickshire in his *The Antiquities of Warwickshire*. Evelyn the diarist records dining with Dugdale when he was eighty-one, when they were guests of the second Earl of Clarendon, Lord Privy Seal, and writes that he had 'his sight and his memory perfect'.   He was a man of wide culture and interests and it is virtually certain that he drew his maps himself.

One of the more colourful characters in British cartography was John Ogilby (1600– 76), who was born either in or near Edinburgh.   He came of a good family, but his father fell on bad times and became a prisoner of the King's Bench.   This meant that John had little education.   But the lad managed to save a little money which he

[1] The above listing is taken from *How to identify Old Maps and Globes* by Raymond Lister (Bell, 1965).

successfully invested in a lottery, consequently obtaining his father's release, and enabling him to bind himself apprentice to a dancing master named Draper. He made quick headway in this career, and had soon saved sufficient money to buy himself out of his apprenticeship. He then began to teach on his own account, and in time had the reputation of being one of the best teachers of the day. As a result he was elected to take part in an important court masque arranged by the Duke of Buckingham, in which he injured and lamed himself. Thereafter, in 1633, Ogilby became attached to the household of the Earl of Strafford, Lord-deputy of Ireland, where he was employed to teach the Earl's children and transcribe papers. He had earlier learned how to use the pike and musket, and he now also became a member of the Earl's troop of guard.

In time he became Deputy-master of the Revels in Ireland, and built a small theatre in St. Werburgh Street, Dublin, which became popular. But ill-luck was soon to overtake him, and in the Civil War of 1641 he lost all his possessions and narrowly escaped with his life. He returned to England, but misfortune had not quite finished with him, for he was shipwrecked on the voyage and arrived here penniless. Ogilby was, however, quite irrepressible, and he went to Cambridge where he learnt Latin from some scholars who took pity on him. Subsequently he made translations of Virgil and Aesop; a little later he learnt Greek and translated Homer. He participated in the arrangements for Charles II's coronation, being solely responsible for what was described as the 'poetical part'. In 1662 he again became Master of the Revels in Ireland and built himself, at a cost of about £2,000, another theatre in Dublin.

Ogilby once more returned to London where he set up business in Whitefriars, just off Fleet Street, as a publisher. Bad luck overtook him again, for his house and stock, valued at £3,000, were destroyed in the Great Fire in 1666. Even this did not deter him, and in partnership with a relative of his wife, William Morgan, he obtained a commission to survey the City of London as 'sworn viewer', and plot out disputed property. They surveyed the whole city and in 1677 published the plan they made from the details.

After rebuilding his house, Ogilby set up again as a publisher and became King's Cosmographer and Geographic Printer. His publications included a series of maps and atlases, of which the most original are his road maps (Plate 38). His *Britannia* of 1675[1] contains the earliest strip road maps of England. From this time the representation

[1] This edition was issued twice, first with seven leaves of text and then with four leaves. A new edition without text followed in the same year; and a repeat in 1698.

of roads on maps became increasingly important, and they were added to those plates by Saxton and Speed that were still being used.   Ogilby was particularly concerned with post roads, and in passing it is interesting to note that it was because of the extension of the postal service outside London in the seventeenth century that the statutory mile of 1760 yards became generally accepted, although it was not finally and officially established until the passing of the Act of Uniformity of Measures in 1824.   Ogilby was also responsible for much of the proliferation of conventional signs that occurred at about this time.   This extract from the preface to his *Britannia* will give some indication of this:

> The *Road* . . . is express'd by double Black Lines if included by Hedges, or Prick'd Lines if open . . . *Capital Towns* are describ'd Ichnographically [i.e. in plan] . . . but the *Lesser Towns* and *Villages*, with the *Mansion Houses, Castles, Churches, Mills, Beacons, Woods, &c.* Sceneographically, or in Prospect.   *Bridges* are usually noted with a Circular Line like an Arch . . . *Rivers* are *Decypher'd* by a treble waved Line or more, and the lesser *Rills* or *Brooks* by a single or double Line, according to their Eminency.

Ogilby's work marked the onset of a period of great cartographical progress that lasted roughly from 1660 to 1760.   In particular there was immense activity on the part of independent British cartographers, engravers and artists who finally shed Netherlandish influence during this period.   There was also a sharp rise in the number of reproductive engravers.

A leading English chart maker of this time was Greenville Collins (*fl.* 1679–93), a captain in the Royal Navy.   He was a Younger Brother of Trinity House and was in 1679 appointed to the command of the frigate, 'Lark'.   Two years later he was given the command of a yacht, 'Merlin', and ordered to chart the coasts of Britain.   This task occupied seven years, and in 1693 Collins's charts were published as *Great Britain's Coasting Pilot*, which contained forty-eight charts on forty-five sheets.   The collection was reissued in 1723, 1738, 1749, 1753, 1756, 1760, 1764 and 1785.   It is a magnificent work, full of all the confident flamboyance of the age.   Many of the charts have elaborate cartouches, some of which are highly appropriate, like that on Chart No. 3 (The English Channel and part of the North Sea), which is a huge cockle shell held up by mermaids, themselves supported by tritons.   It bears the chart's dedication: 'To the Honourable Sr John Ashby Kt. Admiral of the Blew' (Plate 36).   Among other dedicatees is 'Ye Hon[ble] Samuell Pepys Esq[r] Secretary of the Admiralty of England. President of ye Royall Society & Maister of ye Trinity House of Deptford: Stroud'. This appears on the chart of *Harwich Woodbridg and Handfordwater with the Lands from*

39. Bird's-eye map of Devonshire by G. Bickham, from *The British Monarchy*, 1750. *British Museum*.

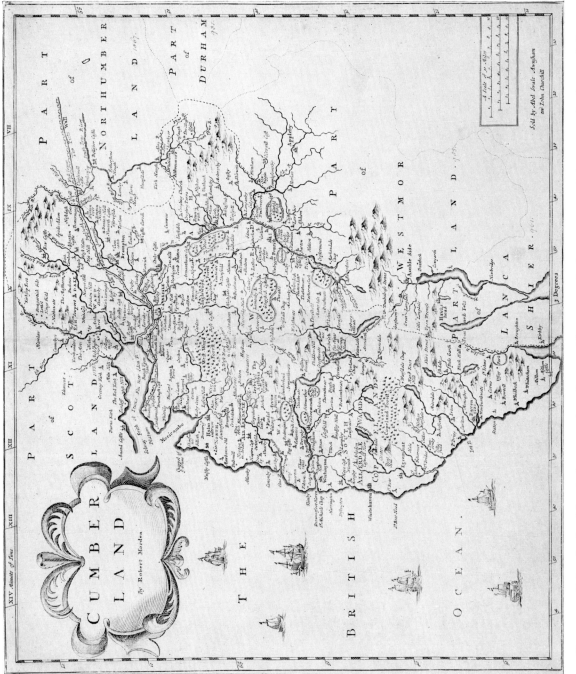

40. Map of Cumberland by Robert Morden, from William Camden's *Britannia*, 1695. *Mr. Richard A. Nicholson.*

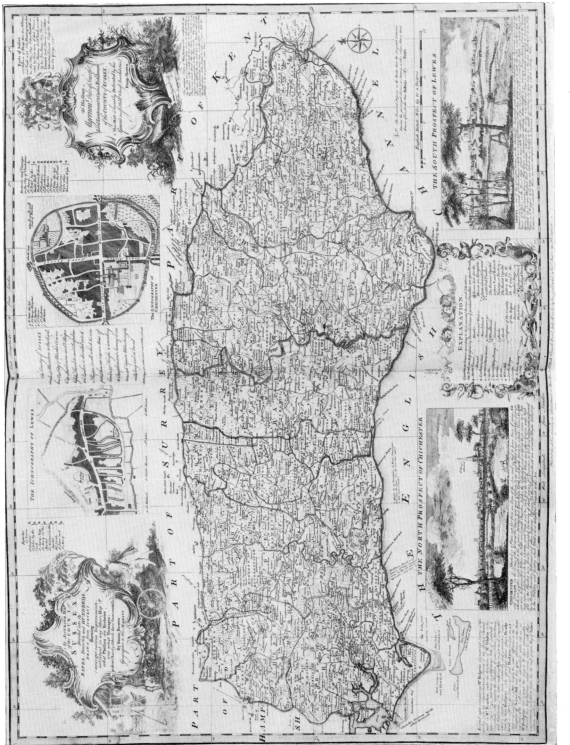

41. Map of Sussex by Emanuel Bowen, *circa* 1760.

British Museum.

42. Map of Berkshire, 1762, by John Rocque. *British Museum.*

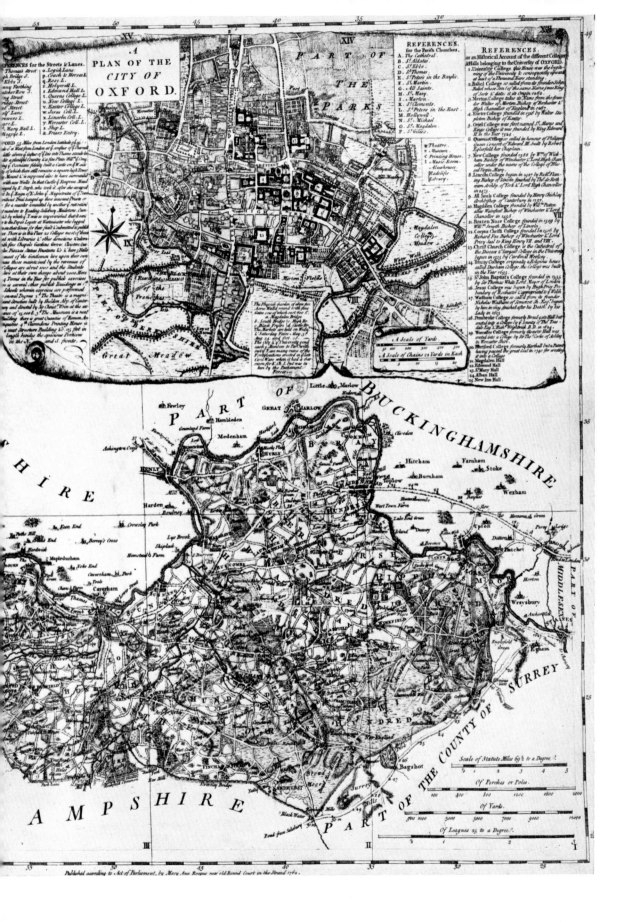

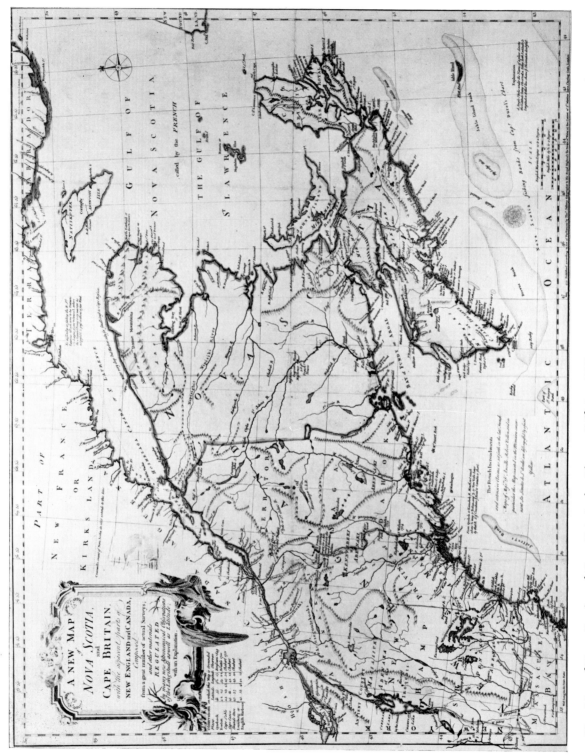

43. Map of Nova Scotia and Cape Britain, by Thomas Jefferies, 1755.

British Museum.

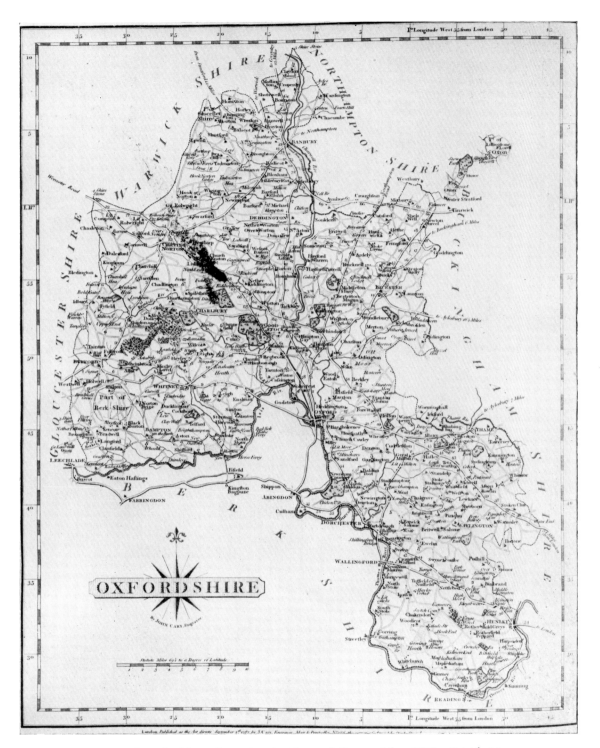

44. Map of Oxfordshire from Cary's *New and Correct Atlas*, 1787. *British Museum.*

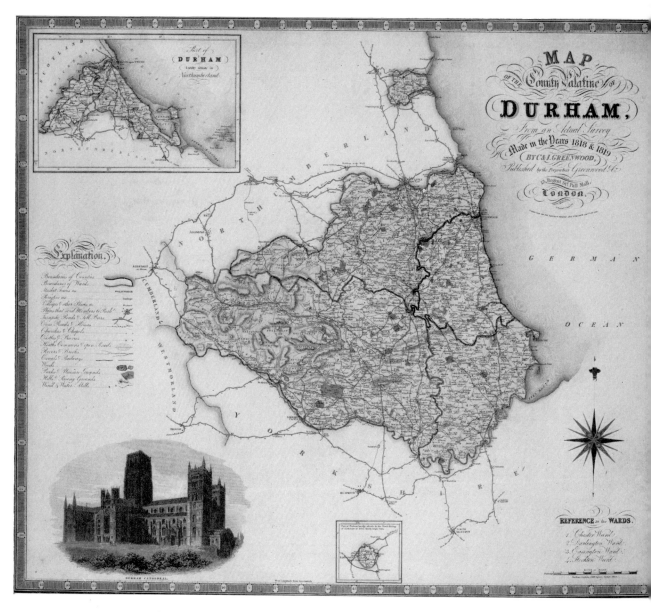

45. Map of Durham by C. and J. Greenwood, 1818–19.   *Mr. Richard A. Nicholson.*

the Nazeland to Hosley Bay, and it is inscribed on a cartouche formed of armorial devices, fishes, putti and navigational instruments. Decoration apart, Collins's charts are notable for their accuracy, being a vast improvement on anything that preceded them. In many ways Collins may be considered as the greatest hydrographer in British cartographic history.

Richard Blome (d. 1705) was a noted publisher of maps in the seventeenth century, though his reputation is tarnished by his business methods. Anthony à Wood described him as being 'esteemed by the chiefest heralds a most impudent person, and the late industrious Garter (Sir W. D[ugdale]) hath told me he gets a livelihood by bold practices'. Wood further stated that among his 'progging tricks' was the employment of impecunious writers to produce works, and then getting contributions from noblemen to promote them. Among his publications were several editions of Guillim's *Display of Heraldrie* (1660–79), *The Fanatick History . . . of the Old Anabaptists and New Quakers* (1660), *A Geographical Description of the four parts of the World, taken from the notes and works of Nicholas Sanson and other eminent travellers and authors* (1670), *A Description of the Island of Jamaica . . .* (1672), *Britannia; or a Geographical Description of the Kingdoms of England, Scotland, and Ireland, with the Isles and Territories thereto belonging; and there is added an Alphabetical Table of the names, titles, and seats of the Nobility and Gentry; illustrated with a Map of each county of England* (1673). *Britannia* carries evidence of Blome's methods of work, for it contains a list of 'Benefactors and promoters of this worke, whose names, titles, seates, and coates of armes, are entered as they gave their encouragements'. Bishop Nicolson the antiquary described this work, with some justice, as a 'most entire piece of theft out of Camden and Speed'. One interesting item in it is a map, engraved by W. Hollar, of London before the Great Fire. It has a decorative border made up of arms of the principal livery companies, some of which are now extinct. It is dedicated to Alderman the Hon. Sir Robert Vyner, and bears his arms (Plate 37). Another work published by Blome was *The Present State of his Majestie's Isles and Territories of America: with new Maps, together with astronomical tables from the year 1686 to 1700* (1687).

Another great map and globe maker of this period was Robert Morden (d. 1703) of London, who first worked about 1668; twenty years later he was in partnership with Thomas Cockerill (*fl.* 1674–1702) at the sign of the Atlas in Cornhill. Though Morden's maps are attractive, with pleasing cartouches and other decorative features, they are unoriginal geographically and are coarsely conceived (Plate 40). Many older conventional symbols are still used, such as 'molehills' for relief, churches for towns, and 'Noah's ark' trees for woodlands. Apparently they were not as successful as Morden would have wished, for he was often short of money, despite the fact that he worked hard. He published many maps and other works, including *Geography rectified; or a description of the world in all its kingdoms . . . illustrated with seventy-six maps* (1680; other editions appeared in 1688, 1693 and 1700); *An Introduction to Astronomy, Geography, Navigation, etc., made easie by the description and uses of the celestial and terrestrial globes* (1702); *The New Description and State of England, containing the Maps of the Counties of England and Wales . . .* (1704); *Pocket-Book Maps of all the Counties of England and Wales* (undated). Morden's separate maps are numerous and include a *Map of the World, drawn according to Mercator's projection* (circa 1700); *A New Map of Ireland* (by R. Morden and J. Overton; *circa* 1680); *The smaller Islands in the British Ocean* (*circa* 1700); *Actuall Survey of London, Westminster, and Southwark* (two sheets; 1700); *A Map containing the towns, villages, gentlemen's houses, roads, rivers . . . for twenty miles round London* (*circa* 1700); *A Mapp of Scotland* ('made by R. Gordon . . . corrected by R. Morden'; *circa* 1700). Morden also made separate maps of English counties, and a number of maps of continental Europe, including maps to illustrate military campaigns. A map of the West Indies was posthumously published in 1740.

Morden's maps were used for an edition of Camden's *Britannia* (1695). This was the first collection of English maps to use the prime meridian of Greenwich throughout. Further editions of *Britannia* with these maps were published in 1715, 1722, 1737, 1753 and 1772.

Emanuel Bowen (d. 1767) was engraver of maps to George II and Louis XV. His publications included a *Complete Atlas of Geography* (1744–7); *English Atlas, with a new set of maps* (*circa* 1745); *Complete Atlas . . . in sixty-eight Maps* (1752); *Atlas Minimus; or a new set of Pocket Maps* (1758). With Thomas Kitchin (1718–84) he published a series of attractive maps of English counties. These were published separately and showed the geographical features in great detail. In free spaces, were views of the principal towns and detailed historical notes. Their cartouches are of rococo design, some containing titles, others dedications, and some also containing vignettes of the county's principal manufactures and trades (Plate 41). Bowen also made separate maps of Persia, Asia Minor and Germany.

Richard Gough in his *British Topography* says Bowen's map of Cambridgeshire is 'very faulty', that of Northamptonshire 'most inaccurate', and those of Derbyshire and Dorsetshire 'incorrect'. Of Dorsetshire he adds 'scarce two places are spelt right'.

Bowen's son, Thomas (d. 1790), was an engraver of charts and maps, including those of the West Indies from the surveys of Captain James Speer; a *New Projection of the Eastern and Western Hemispheres of the Earth* (1776); *Accurate Map of the Russian Empire in Europe and Asia* (1778); and maps of the road from London to St. Davids and of the country twenty miles around London (*circa* 1750). He also carried out work for *The Survey and Maps of the Roads in North Britain* (1776) by George Taylor and Andrew Skinner. Thomas Bowen is said to have been very old when he died in 1790 in the workhouse at Clerkenwell.

A return to the bird's-eye view type of map or semi-map was made by George Bickham (1684–1771) of Covent Garden, London, who engraved a series of forty-eight English counties with the towns marked on them, and with figures in the foreground wearing dress typical of the district (Plate 39). They were first published in a complete work in 1754 under the title of *The British Monarchy*. There is, however, a title page dated 1743, and the views themselves are dated from 1750 to 1754. In 1796 they were published by Laurie and Whittle in a new edition under the title *A Curious Antique Collection of Birds-eye Views of the several counties in England and Wales*. Bickham's maps are scarce today.

With one or two exceptions, the works of the cartographers so far discussed are notable more for their decorative qualities and in some cases their accuracy, than for their geographical originality. Most of their work followed geographical practices that had obtained for many years on the Continent. In the eighteenth century, however, and in particular towards its end, British cartography made great advances. This resulted largely from the country's expanding maritime influence, particularly in India and North America. Typical examples of maps produced at this time are Major James Rennell's *Bengal Atlas* (1779) and *Map of Hindoustan* (1782), and John Mitchell's *Map of the British and French Dominions in North America* (1755; published by Thomas Jefferys).

Between 1734 and 1762 over one hundred maps and related documents were published by the surveyor, publisher and engraver, John Rocque (*circa* 1704/5–1762), a Huguenot who settled in England and became naturalised. His brother Bartholomew, a landscape gardener, followed him here. Apart from original works Rocque published reproductions of existing maps; he also sold work which was entirely that of other cartographers and engravers.

Among those who worked for Rocque were Kitchin, John Pine, Richard Parr and John Tinney. He also employed foreign engravers from time to time; a letter exists in which it is said that he employed ten foreign draughtsmen and engravers.

A map of Bristol with two views published in 1743 was Rocque's first real cartographical exercise.   But while his publications before that date dealt almost entirely
with plans and views of country houses and gardens, he had since 1737 been surveying
for a map of London and in that year also began carrying out surveys, which continued
until 1751, for his series of large-scale county maps (Plate 42).   The sale of these and
other maps prospered, and about 1751 he moved from his quarters in the vicinity of
Hyde Park and St. James's to larger premises in the Strand, where, apart from a
temporary stay in Dublin sometime between 1753 and 1756, he remained until his
death.

Rocque's *New and Accurate Map of London* appeared in 1744.   Another of his works—
A List of the works of John Rocque, Chorographer to H.R.H. the Prince of Wales—was
issued about 1755, and contains about seventy items, including 'A great variety of
foreign maps, plans, battles, sieges, a complete Atlas', and 'A set of maps of all different
parts of the world at 1/6 each'.   He issued other lists of works besides these, and hand-
bills and prospectuses giving details of forthcoming publications.

Rocque's output was considerable, but his most original and important works remain
his city plans and his large-scale maps of London, Berkshire, Middlesex, Shropshire and
Surrey.   All these city plans and maps are accurate and original.   It is worthy of
comment that he used his own conventional signs, some of which are remarkably
modern.   In his map of London, for example, he shows orchards and woods somewhat
as they are shown in Ordnance Survey maps.   He also differentiated the various kinds
of land—grass, pasture or arable.

Here is a brief selection of Rocque's maps and plans.   *Plan of house, gardens, park and
hermitage of their Majesties, with views and account of Richmond Palace* (1734); plan of
gardens and house at Weybridge (1737); plan of Bristol in eight sheets (1743; engraved
by Pine); plan of Berlin in four sheets (1745); plan of Paris (1748 and 1754); *The Small
British Atlas* (1753); map of Middlesex in four sheets (1754), and in one sheet (1757);
map of the British and French possessions in North America (1755; engraved by
Thomas Kitchin); plan of Dublin in four sheets (1756); plan of the city of Trichinopoly
and surrounding country (1757); *Traveller's Companion*, post-roads of England and
Wales (1763 and various subsequent editions).

An important if little-known cartographer of this period was Henry Beighton
(1687–1743), whose output consisted of maps of Warwickshire.   Despite this limitation
his maps are notable for their clarity and accuracy, in which respect he surpasses many
better-known cartographers.   He was born at Chilvers Coton, near Nuneaton, of an

old yeoman family.   By the time he was twenty-four he was planning a new large-scale survey of Warwickshire, and although this came to nothing then, he did realise his ambition a few years later.   He was keenly interested in all kinds of subjects.   For a long time he collected data on weather, which he sent to the Royal Society; he became editor of *The Ladies' Diary; or, The Woman's Almanack*; he lectured; he studied the Newcomen steam engine, for which he invented a new valve, and may have known Thomas Newcomen himself.   Yet manifold interests and activities (and the foregoing list does not exhaust them) did not prevent Beighton from producing some of the best county maps so far seen in this country.   He had a sound business sense and offered to include the arms of local gentlemen in his maps for half-a-guinea each.   Beighton did not hesitate to blow his own trumpet and to denigrate the reputations of his predecessors.   In referring to the use of mathematics in surveying, he said that most contemporary surveyors were 'Lame, Deficient or Trifling on that Head'; but he went too far in claiming that 'most of our County maps are only Transcripts from . . . *Saxton*, or *Nordens*, made before 1576'.

Thomas Jefferys (1695?–1771) was an important figure in English cartography and occupied the post of 'Geographer of Maps and Charts' to the Prince of Wales, afterwards George III.   He took a leading part in new topographical surveys of the United Kingdom which by the end of the century had transformed the mapping of the country. Where overseas maps were concerned he was particularly noted for his charts of American coastal areas and for his maps of British colonies.   His American coastal charts were based on observations made on the spot by sailors.

Little is known of Jefferys's early life.   He was probably apprenticed to an engraver, for he seems to have considered himself as an engraver first and a geographer second. He did a good deal of engraving apart from that of maps.   He made book illustrations, views of buildings and estates, and what he described as 'a great Variety of Italian, French and Chinese Ornaments for Invitation Cards, Visiting Tickets, and Cards of Thanks'.   He published his first map in 1732; it was *An Accurate Plan of the Citties of London and Westminster and the Borough of Southwark*.   It was not an original work, being a re-working of an older plate by William Morgan (1681–2).   Even his own maps were not at first very original, being based on contemporary sources.   Such was *The Small English Atlas* (1749).   But original work followed, like the *New Map of Nova Scotia, and Cape Britain* (1755; Plate 43), which was published simultaneously with John Green's *Explanation For the New Map of Novia Scotia and Cape Britain*.   Like most of Jefferys's maps the *New Map of Nova Scotia* is plain and workmanlike, concentrating on an accurate presentation of geographical data.   But it is not without decorative features, particularly in its restrained rococo cartouche, showing a canoe and rapids. Jefferys's maps and charts of North America became so well known that American publishers, when seeking subscriptions for new maps, would advertise that 'Mr Thomas Jefferys, a skilful geographer in London, will be the engraver'.

Jefferys's first business premises were at Clerkenwell.   In 1750 he married and moved into another shop at 5 Charing Cross, which was his home, workshop and retail shop.   His output and trade were prolific, but he nevertheless was bankrupt in 1766. The reasons for his bankruptcy are not clear, but it is thought that he may have overreached himself in paying for county surveys, hoping by these to realise his ambition to win a Society of Arts award.   That he worked for such an award we have on the authority of Richard Gough who, writing of Jefferys's map of Yorkshire, said: 'Mr Jefferys undertook this and other such surveys in consequence of a premium of 100 l. offered by the Society of Arts for a county map . . . It is . . . the best map of the county that has been made.' Nevertheless, Jefferys never received the award.

Whatever the cause of his bankruptcy, Jefferys was in business again by 1767, 'having', as he said, 'found some Friends who have been compassionate enough to re-instate me in my Shop'.   Among these friends was probably Robert Sayer (*fl.* 1780–1810), with whom Jefferys seems to have been in association or partnership.   Their names appear together on some maps which bear the imprint 'Robt. Sayer in Fleet Street & Thos. Jefferys in the Strand'.

Jefferys died in 1771 and was succeeded by William Faden (b. 1750), who may have entered into partnership with him before this.   Faden's 1822 catalogue lists over three hundred and fifty cartographic items—globes, maps, plans, charts and atlases.   The atlases were made up from sheets to suit customers' individual requirements, and manuscript contents lists were inserted.   In 1777 Faden published a North American atlas of thirty-four maps.   He also published some important North American plans, including those of New York and Philadelphia, and of the military operations of the War of Independence.   The first sheets and maps of the Ordnance Survey were engraved in Faden's workshops at the end of the eighteenth century, before the Survey had its own premises and staff.

John Green (d. 1757) has been mentioned as the author of the *Explanation for the New Map of Nova Scotia*.   In fact he was for about five years closely associated with Jefferys as his geographical editor, both under his own name and under the pseudonym

Braddock Mead. Green was a retiring character and more often than not signed his work with initials only. He contributed much to the increasing degree of accuracy evident in English eighteenth-century maps. He placed a list of authorities on each of his maps by which their accuracy might be assessed, and he distinguished, by a line drawn under their names, places whose position had been determined astronomically. Moreover, he supplied explanatory tables of geographical terms that appeared on maps of foreign parts. He endeavoured, too, to solve the problem of the transliteration of foreign names into English. In short, in many ways he anticipated modern carto-graphical methods.

Some of Green's works are *A Chart of North and South America, including the Atlantic and Pacific Oceans, with the nearest Coasts of Europe, Africa and Asia* (six sheets; 1753. Published by Jefferys; republished in 1775 by R. Sayer and J. Bennett); *A Map of the most inhabited part of New England, containing the Provinces of Massachusets Bay and New Hampshire, with the Colonies of Konektikut and Rhode Island, Divided into Counties and Townships: The whole composed from Actual Surveys and its Situation adjusted by Astro-nomical Observations* (four sheets; 1744 and 1755. Published by Jefferys; republished in 1794 by Laurie and Whittle). Green was also author of *Remarks, In Support of the New Chart of North and South America* (Jeffreys, 1753), which was related to the first-named map above; *A Journey from Aleppo to Damascus . . .* (1736), which contains a map of the road from Aleppo to Damascus with the adjacent parts of Syria; and an English version of P. J. B. Du Halde's *A Description of the Empire of China and Chinese Tartary* (1741), which also contained maps.

Another cartographer who relied upon geographical verisimilitude rather than on decoration was John Cary (*circa* 1754–1835). But his maps have a chaste beauty, a Quaker-like plainness that is decoration enough for some collectors (Plate 44). He issued several atlases at the end of the eighteenth and beginning of the nineteenth centuries, including a *New Universal Atlas* (1808 and subsequent editions); *New and Correct English Atlas* (1787 and subsequent editions; one, as late as 1862, was published by George Frederick Cruchley); *Cary's English Atlas* (1809 and subsequent editions); *New British Atlas* (1805; published in collaboration with John Stockdale); and *Cary's Travellers' Companion* (1790 and subsequent editions). Cary's brother William (1759–1825), a pupil of Jesse Ramsden, the optician and mechanician, was a well-known maker of globes and instruments, and had his own business by 1790. John, too, made globes but was better known for his maps.

John Cary engraved his maps until about 1800, but after that he employed others to carry out his engraving, colouring and drawing, while he concentrated on the management of the business. His firm was known as G. and J. Cary, the initials being those of his sons George and John II; he is thought to have transferred the business to them while remaining an active partner. A wide variety of publications and goods was handled by the firm, including maps, plans, road-books, celestial charts, geological maps, guides, orreries and magic lanterns. Prices varied considerably, from a guinea for a four-sheet world map measuring 5 feet 10 inches by 3 feet 2 inches, which could be had mounted on wooden rollers for an extra fourteen shillings, to a *General Atlas of the World*, bound and coloured, for £10 10s. 0d. Cary's work also included the supervision of the Postmaster-General's survey of nine thousand miles of British turnpike roads. His many-sided output made him one of the greatest influences on British cartography since Christopher Saxton.

The Cary business closed down about 1850, and soon afterwards the plates were acquired by George Frederick Cruchley (*fl.* 1822–75). Cruchley's name appears on the maps thereafter, usually with suitable acknowledgements to Cary. In 1876 Cruchley's plates were acquired by Messrs. Gall and Inglis, a firm that still exists as publishers in London. The maps were still being issued by them at the beginning of the present century, with additions, as cycling and motoring maps.

Another excellent cartographer of the early nineteenth century was Charles Smith (*fl.* 1800–52) whose *New English Atlas* went into many editions from 1808 to 1864.

Aaron Arrowsmith (1750–1823) was one of the foremost of late eighteenth-century British cartographers, and the founder of a dynasty of mapmakers. He was born at Winston in Durham. His father died when he was young, and his mother remarried, to a man who wasted all the money the boy's father had left. He was thus not only forced at a tender age to work for his living, but received very little education. Aaron came to London about 1770, where he was employed by John Cary—some authorities claiming that he took all the pedometer measurements for Cary's *Actual Survey of the Great Post Roads* (1784). By 1790 he was established in his own business in Long Acre, from which address on 1 April in that year he published his first map, now of great rarity: *A Chart of the World upon Mercator's Projection, showing all the New Discoveries . . . with the Tracts of the most distinguished Navigators since 1700.* Many other maps and surveys followed this, all of them as popular abroad as they were in England. Among them are *Map of North America* (1796); *Map of Scotland, constructed from original materials obtained by the authority of the Parliamentary Commissioners for Making Roads and Bridges in the Highlands* (four sheets; 1807); *Atlas of Southern India* (eighteen sheets; 1822); and

Pilot from England to Canton (seven charts on twenty-three sheets; 1806). There were also maps of the United States (1796, 1802, 1815, 1819), Panama Harbour (1806), West Indies (1803, 1810), Africa (four sheets; 1802, 1811), Asia (1801), Persia (1813), and Egypt (1802, 1807).

Some of Arrowsmith's maps are large. His Pacific Chart (1798), for example, was issued in nine sheets and measured 6 feet by 7 feet 6 inches. They provide valuable historical source material, for they are clear, correct and well engraved.

Arrowsmith moved from Long Acre to Rathbone Place in 1802. In 1814 he removed to Soho Square, where he died on 23 April 1823. After his death his sons Aaron and Samuel, and later his nephew John, carried on his business, though they continued to use his name. The business operated until Samuel's death in 1839. Works published by the younger Arrowsmiths include *Atlas of Ancient Geography* (1829); *Atlas of Modern Geography* (1830); and *Bible Atlas* (1835). Apart from their cartouches the Arrowsmith maps are plain and unornamental (Plate 48). Like Cary's maps, while not being decorative in intention, they derive decorative value from their very restraint.

John Arrowsmith (1790–1873), the nephew of Aaron, came to London in 1810. He at first worked for his uncle, but set up on his own after Aaron's death in 1823. Then some time later he returned to the business once again. His first publication was a *London Atlas* (1834) which passed through several editions. In the preface to the 1858 edition he says he had 'examined more than 10,000 sheets of private maps, charts and plans, thereby rectifying all the labours of his predecessors'. He published large maps of India (twenty sheets), England and Wales (eighteen sheets), Spain (twelve sheets), Pacific Ocean (nine sheets), and the World (ten sheets). The following were all published in eight sheets apiece: Atlantic Ocean, British Channel, Canada and Ceylon. The following in six sheets each: America, Australia, France, Germany and Thebes. The following each in four sheets: Africa, America, Asia, Bolivia, East Indies, West Indies and Italy. In addition he published numerous smaller maps.

John Arrowsmith's maps of Australia are of particular importance, for he knew many of the Australian explorers and was able to use much of the information they gave him. He retired in 1861, but continued to improve some of his maps. He had been a Founding Fellow of the Royal Geographical Society in 1830, and the Society presented him with its gold medal in 1863 for his services to geography.

Another family cartographical business was that of the Greenwood brothers, Christopher (sometimes mistakenly called Charles) and John. Christopher was born at Gisburn in Yorkshire in 1786 and died at Hackney, London, in 1855.   He started in business in Wakefield as a surveyor, but came to London in 1818 where he set up in business in Leicester Square.   Two years later, in 1820, he entered into a partnership with George Pringle Senr. and his son, another George.   In the following year their business became known as Greenwood and Co., county surveyors.   Their office was at 70 Queen Street, Cheapside, London.   Later still they moved to Lower Regent Street, by which time Christopher's younger brother John had become a partner.   After this move their fortunes declined and although Christopher tried hard to re-establish himself, his mapmaking activities ended in the 1840's.   Before this, in 1838, John had left the firm and had set himself up as a surveyor in his home town of Gisburn.

One of Christopher Greenwood's foremost productions was his *Atlas of the Counties of England* (1834), decoratively one of the most ambitious cartographical productions of the nineteenth century.   Its engraving is excellent, and each map is provided with a view of a noted building in the county (Plate 45).   He also issued a series of separate one-inch county maps, including Bedfordshire (four sheets; 1826), Cornwall (six sheets; 1827), Northamptonshire (four sheets; 1826), and Yorkshire (nine sheets; 1817–18). The plate of the last-named map was purchased in 1827 by the London cartographical publishers, Henry Teesdale and C. Stocking, who published a lithographic copy without acknowledging Greenwood.   A Fleet Street map-mounter, Edward Ruff, likewise purchased the plate of Greenwood's map of Warwickshire and impressions taken from this were also issued without acknowledgement.

A rival of the Greenwoods was A. Bryant (*fl.* 1822–35) of Great Ormond Street, London.   He published twelve county maps and one of the East Riding of Yorkshire. Another rival was Thomas Hodgson of Lancaster.   Greenwood carried on an acrimonious public quarrel with Hodgson in the columns of the *Westmorland Advertiser*. Both participants were cut down to size by another correspondent, who declared that, 'Instead of their ridiculous squabbles it is high time for them to attend to their proper business.'

I have just mentioned a lithographed copy of one of Greenwood's maps.   It was during the first half of the nineteenth century that lithography came into use for map production.   The process, which was invented in 1796 by Aloys Senefelder, is based on the antagonism of water for grease.   A perfectly flat stone surface is made chemically clean, and thus very sensitive to grease.   The design is drawn on its surface, or transferred from a copper plate, by oil-based lithographic ink (or, in the case of much art work, chalk made of grease).   Those parts of the stone free of drawn lines and areas are then desensitised by the application of dilute nitric acid and gum arabic. The stone is moistened, and when an inked roller is applied to it the ink will adhere to the drawn parts, but not the other areas. Impressions are then taken by applying paper, in a press, to the inked surface.

Lithography became one of the main methods of map production from this time on, and some of the most attractive nineteenth-century maps were produced in this way. Foremost among these are the county maps of Thomas Moule (1784–1851), which were first published in *The English Counties Delineated; or a Topographical Description of England. Illustrated by a Map of London and a complete Series of County Maps* (two volumes; 1836–41). There were various editions of these maps, which were later issued in *Barclay's Complete and Universal Dictionary* (1842 and subsequent editions). The maps are considerably detailed, and show roads and railways. Artistically they belong to the Gothic revival, and are richly embellished with decorative borders, armorial bearings, views and figures. They are as expressive of the spirit of the Victorian Age as are Saxton's of that of Elizabeth (Plate 47).

Moule was not a cartographer in the usually accepted sense, but a scholar and a writer on antiquities and heraldry. He was born in Marylebone, London. From 1816 to 1823 or thereabouts he made his living as a bookseller in Duke Street, Grosvenor Square. Later he worked at the General Post Office as inspector of 'blind' letters—those letters which are badly and illegibly addressed. He was also a chamber-keeper in the department of the Lord Chamberlain, and this office provided him with a residence in the Stable Yard, St. James's Palace. Yet, though Moule was not a typical cartographer, he was a conscientious one, for when he was compiling his set of county maps he visited every county except Devon and Cornwall. Moule's other works included *A Table of Dates for the use of Genealogists and Antiquaries* (1820); *Antiquities in Westminster Abbey, illustrated by twelve plates, from drawings by G. P. Harding* (1825); *An Essay on the Roman Villas of the Augustan Age . . . and on the Remains of Roman Domestic Edifices discovered in Great Britain* (1833); and *Heraldry of Fish, Notices of the principal families bearing Fish in their Arms* (1842).

But as far as decorative maps were concerned, Moule's could be described as a swan song. In the main, Victorian cartography was plain and practical rather than decorative, and became the concern of government departments and institutions rather than of individuals. Typical was the Ordnance Survey, known at first as the Trigonometrical Survey. This was the concern of the Board of Ordnance and was undertaken largely for military purposes, after it had been realised, during the campaign which ended with the defeat of the Young Pretender at Culloden in 1746, just how unreliable military maps were.   The Survey was founded in 1791, but the whole of Great Britain was not covered by it until 1870.

Some of these plain Victorian maps are very attractive and interesting.   There is, for example, the small and rare group of telegraphic maps, like Henry George Collins's *Railway and Telegraphic Map of Yorkshire* (1858), George Frederick Cruchley's *Railway and Telegraph County Maps showing all the Railways and Stations also the Telegraphic Lines and Stations* (*circa* 1860), and *The Electric Telegraph Company's Map of the Telegraph Lines of Europe* (1854; lithographed by Day and Son; Plate 53).   These maps recall the atmosphere of nineteenth-century commercial enterprise as vividly as Michael Drayton's maps recall that of the world of seventeenth-century poetry.   School atlases are another group, and one in which a collector can still find some genuine bargains. There are many of them, but by way of an example we may mention *A New Royal Atlas* (1810) by the Rev. John Evans (1767–1827) and Walter McLeod's *A Hand Atlas for Class Teaching* (1858; Plate 46), in both of which it is interesting to note 'molehills' still being used to indicate relief.   Such children's books as the Rev. Isaac Taylor's *Scenes in Europe* (1818), *Scenes in Asia* (1819), *Scenes in Africa* (1820) and *Scenes in America* (1821) contain neat little folding maps, sometimes hand-coloured.

There are, too, railway maps, canal maps, and other types of specialised maps such as the hunting maps of J. and C. Walker.   Even sepulchral maps were proposed by William Godwin, Shelley's father-in-law, in his *Essay on Sepulchres* published in 1809 at the height of the Romantic Movement, in which he said: 'I spoke a while ago of maps, in which the scenes of famous battles were distinguished with a peculiar mark. Why might not something of this kind be introduced in the subject before us?   It might be called, the Atlas of those who Have Lived, for the Use of Men Hereafter to be Born.   It might be plentifully marked with meridian lines and circles of latitude, "with centric and eccentric scribbled o'er", so as to ascertain with incredible minuteness where the monuments of eminent men had been, and where their ashes continue to repose.'

Above all the nineteenth-century map was practical.   It was put to all kinds of uses new and old, even to the suppression of crime.   Major-General Sir William Sleeman (1788–1856), who headed a government department for the suppression of the religious secret society of Thugs in India, used a large wall-map showing the Thugs' villages, the location of their crimes, and their burial grounds.   This revealed a certain pattern which helped him in his campaign.   This seems a long way from the wind-swept coastlines of Saxton, from the shires of Speed, or from Bickham's pastoral counties. But all, whether the decorative maps of Saxton and Speed, or the purely utilitarian maps of the later Victorians, are part of our cartographic heritage and an expression of our history.

BIBLIOGRAPHY

CHUBB, THOMAS: *A descriptive list of the printed maps of Somersetshire, 1575–1914.* 1914.

—— *The printed maps in the atlases of Great Britain and Ireland. A bibliography.* 1927.

—— and STEPHEN, G. H.: *Descriptive list of the printed maps of Norfolk, 1574–1916: Descriptive list of Norwich plans 1541–1914.* 1928.

CLOSE, SIR CHARLES: *The early years of the Ordnance Survey.* 1926.

CRONE, G. R.: *Early maps of the British Isles A.D. 1000–1579.* 1961.

—— 'Early Atlases of the British Isles'. *Book Handbook* No. 6. 1948.

—— 'John Green. Notes on a neglected eighteenth century geographer and cartographer'. *Imago Mundi*, Vol. VI, pp. 85–91. 1949.

—— *Maps and their Makers.* 1968.

DAY, ARCHIBALD: *The Admiralty Hydrographic Service, 1795–1919.* 1968.

DRAYTON, MICHAEL: *Poly-Olbion.* Edited by J. William Hebel. 1933.

EMMISON, F. G.: *Catalogue of maps in the Essex record office 1566–1860.* 1947.

FORDHAM, SIR HERBERT GEORGE: *Hand-list of catalogues and works of reference relating to carto-bibliography and kindred subjects for Great Britain and Ireland 1720 to 1927.* 1928.

—— *John Cary.* 1910 and 1925.

—— *Notes on British and Irish itineraries and road books.* 1912.

—— *Notes on the cartography of the counties of England and Wales.* 1908.

—— *Road-books and itineraries of Great Britain.* 1924.

—— *Road-books and itineraries of Ireland 1647–1850.* 1923.

—— *Some notable surveyors and map-makers of the 16th, 17th and 18th centuries and their work.* 1929.

—— *Studies in carto-bibliography, British and French.* 1914.

GERISH, W. B.: *The Hertfordshire historian, John Norden, 1548–1626 (?): a biography.* 1903.

GOUGH, R.: *British topography* (2 vols). 1780.

GRACE, F.: *Catalogue of maps, plans and views of London.* 1878.

HALLIDAY, F. E. (editor): *Richard Carew . . . The survey of Cornwall.* With John Norden's maps. 1953.

HARLEY, J. B.: 'The bankruptcy of Thomas Jefferies; an episode in the economic history of eighteenth century map-making'. *Imago Mundi*, Vol. XX, pp. 27–48. 1966.

HARLEY, J. B.: *Christopher Greenwood county map-maker and his Worcestershire map of 1822.* 1962.

—— *The First Edition of the One Inch Ordnance Survey*, a reprint in 97 sheets: commenced 1969.

—— 'The re-mapping of England 1750–1800'. *Imago Mundi*, Vol. XIX, pp. 56–7. 1965.

HARVEY, P. D. A. and THORPE, H.: *The printed maps of Warwickshire, 1576–1900.* 1959.

HEAWOOD, EDWARD: *English county maps in the Collection of the Royal Geographical Society.* 1932.

—— 'John Adams and his Map of England'. *Geographical Journal*, Vol. LXXIX. 1932.

HUMPHREYS, A. L.: *A hand-book to county bibliography.* 1927.

INGLIS, H. R. G. (and others): *The early maps of Scotland.* 1936.

LYNAM, EDWARD: *British maps and map-makers.* 1944.

—— 'English Maps and Map-makers of the 16th Century'. *Geographical Journal*, Vol. CXVI. 1950.

—— *The map of the British Isles of 1546.* 1934.

—— *The mapmaker's art.* 1953.

—— *Middle Level of the fens and its reclamation . . . with maps of Fenland.* 1936.

MITTON, E. E.: *Maps of Old London.* 1908.

NORTH, F. J.: 'Humphrey Lhuyd's Maps of England and Wales'. *Archaeologia Cambrensis.* 1937.

—— *The Map of Wales before 1600 A.D.* 1935.

—— *Maps, their history and uses, with special reference to Wales.* 1933.

PALMER, CAPT. H. S.: *The Ordnance Survey of the kingdom.* 1873.

POLLARD, A. W.: 'The Unity of John Norden'. *The Library* New Series, Vol. VI. 1926.

PRINCE, HUGH: *Parks in England.* 1967.

ROBINSON, A. H. W.: *Marine cartography in Britain. A history of the sea chart to 1855.* 1962.

RODGER, ELIZABETH M.: *The large-scale county maps of the British Isles.* 1960.

SANFORD, W. G.: *The Sussex Scene in books and maps.* 1951.

SHEERER, J. E.: *Old maps and map-makers of Scotland.* 1905.

SKELTON, R. A.: 'Pieter van den Keere'. *The Library*, Fifth Series, Vol. V. 1950.

SPEED, JOHN: *John Speed's England* (4 vols). 1953. Edited by John Arlott.

TAYLOR, E. G. R. 'Notes on John Adams and contemporary Map-Makers'. *Geographical Journal*, Vol. XCVII. 1941.

—— 'Robert Hooke and the cartographic Projects of the late 17th Century'. *Geographical Journal*, Vol. XC. 1941.

TOOLEY, R. V.: *Late Tudor and Early Stuart Geography 1583–1650.* 1934.

—— *Tudor Geography 1485–1583.* 1935.

—— *Maps and map-makers.* 1962.

VARLEY, JOHN: 'John Rocque. Engraver, Surveyor, Cartographer and Map-seller'. *Imago Mundi.* Vol. V, pp. 83–91. 1948.

WATERS, D. W.: *The Rutters of the Sea . . . A study of the first English and French printed sailing directions . . .* 1967.

WHITAKER, H.: *The Harold Whitaker collection of county atlases, road-books and maps presented to the University of Leeds. A Catalogue.* 1947.

—— 'The Later editions of Saxton's maps'. *Imago Mundi*, Vol. III, pp. 72–86. 1939.

WILSON, E.: *The Story of the Blue-back Chart.* 1937.

CHAPTER SIX

Scandinavia and Russia

W<small>E</small> have now dealt with the cartography of the main map-producing countries of Europe, but Scandinavia and Russia still remain to be discussed.

Scandinavia had appeared on maps since the time of Ptolemy, who was the first geographer in the European tradition to show it, but the first separate map of the area was made by a Dane, Claus Claussen Svart, who lived for a time at Rome, and is better known by his Latinised name of Claudius Clavus. This map, which was drawn in 1424, was the first post-mediaeval map to be added to Ptolemy's *Geography* and was reproduced both by woodcut and copper engraving. It had wide influence in the fifteenth and early sixteenth centuries.

Other early Scandinavian maps include one in Hartmann Schedel's *Liber Cronicarum* or *Nuremberg Chronicle* (1493). This is a wood-cut map of central Europe, but it includes also Southern Scandinavia, the Baltic and Denmark. It was engraved after an original by Cardinal Nicolas Cusanus, which was first printed in 1491. There was also a map of the area by the Bavarian astronomer, Jacob Ziegler, which was the first printed map to show Finland. It was printed in Vienna in 1532 in Ziegler's book on the Holy Land and surrounding area *Quae intus continentur.* Information on which Ziegler based this map was supplied by two Swedes, Johann Magnus, Archbishop of Uppsala, and Erik Walkendorf, Archbishop of Drontheim (Trondheim). It was used as source material by other cartographers including Gastaldi, who used it in his 1548 edition of Ptolemy, and by Sebastian Münster in his *Cosmographia.* A more detailed map that included Scandinavia was published in 1539, the *Carta Marina et descriptio septentrionalium terrarum* of Olaus Magnus (1490–1558), a Swede of Uppsala. It was printed in Venice from woodblocks on nine sheets, and was accompanied by a descriptive text *Ain kurze Auslegung der neuen Mappen von den alten Goettenreich und anderen Nordlenden.*

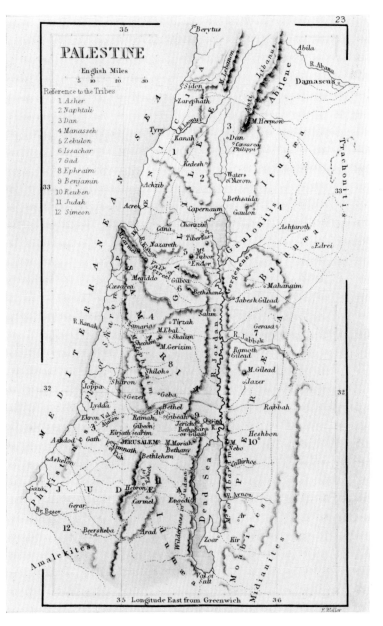

46. Map of Palestine from *A Hand-Atlas for Class Teaching* by Walter McLeod, 1858. Engraved by E. Weller. *Private Collection.*

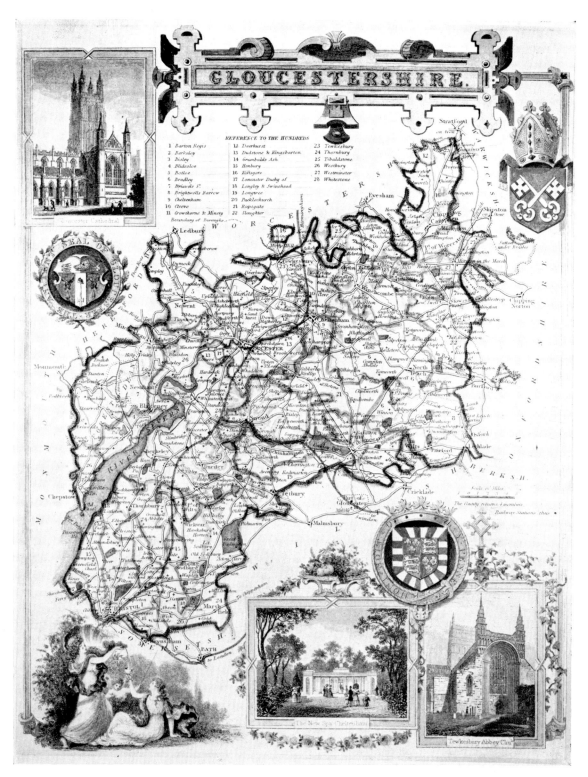

47.   Map of Gloucestershire by T. Moule, 1836–9.   *Private Collection.*

48.   Map of Anglesey, Caernarvon etc. by Aaron Arrowsmith, 1818.   *British Museum.*

49. Map of Denmark from Hogenberg's *Civitates Orbis Terrarum*, 1573–1618.

British Museum.

50. Lyatsky's map of Muscovy as produced by F. Hogenberg, 1570.

British Museum.

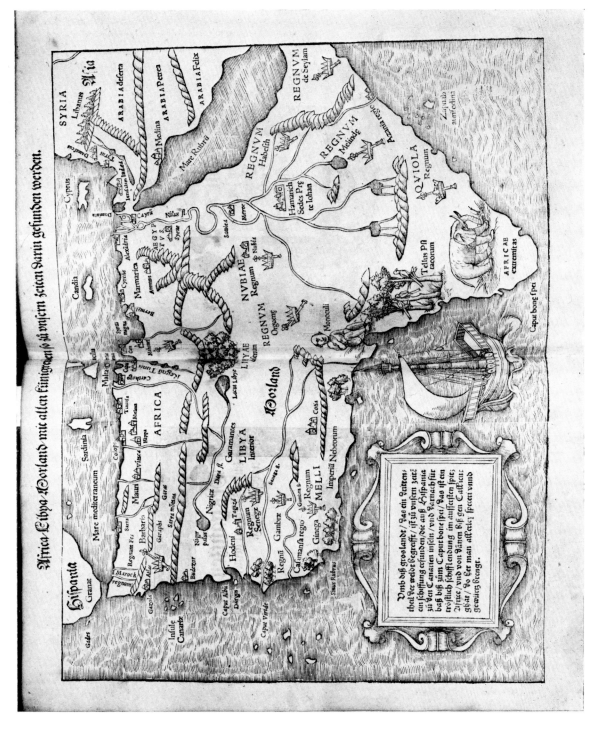

51. Woodcut map of Africa from Sebastian Münster's *Cosmographia*, 1544.

British Museum.

52. *Moscoviæ maximi amplissimique* by De Jode, 1570.   *British Museum.*

The *Carta Marina* depicted Denmark, Finland, Norway, Sweden and Iceland, and was decorated with engravings of Scandinavian costumes and animals. It is of the greatest rarity, only one impression having survived, which is now in the Munich State Library. It was the most influential map of Northern Europe in the sixteenth century. An engraved re-issue of 1572 by Lafreri is almost as rare, only seven impressions being recorded.

Yet another rare map of Scandinavia was published by Cornelis Anthonisz in 1543: his woodcut *Caerte van Oostlandt* on nine sheets. No copy of the first issue remains, and only one copy of the second. Like the *Carta Marina*, the map by Anthonisz had widespread influence and was copied by Michaelis Tramezini (1558), Ortelius (1570) and Mercator (1595). An engraved map of Scandinavia on nine sheets, entitled *Terrarum Septentrionalium*, was published at Antwerp in 1562 by Lieven Algoet. It was used by De Jode in his atlases of 1578 and 1593. There was also a map of the area by the Zeno brothers (1558), which was copied from that of Olaus Magnus.

Mark Jorden (*circa* 1531–95) was an outstanding Danish cartographer of the period. He published a map of Denmark in 1552, but no copy is now known, although one copy remains of his woodcut map of Holstein, which was published in 1559 at Hamburg. Another map of Denmark by Jorden was published in 1585 in Volume IV of Braun and Hogenberg's *Civitates Orbis Terrarum* (Plate 49). It is the only map among that large collection of town plans, some of which, however, are of Scandinavian cities, such as Bergen, Copenhagen and Stockholm. Jansson re-issued these plans in his 1657 *Theatrum Urbium*, with the addition of others, including Elsenor (Helsingor) and Ripen, and Visbia (Visby). Johann Honter (*fl.* 1540) issued a miniature atlas which showed the Danish islands, while a fine Dutch map of Scandinavia engraved by Adriaen Veen (1572–1613) was issued at Amsterdam in 1613.

One of the greatest influences in northern cartography was the Danish astronomer Tycho Brahe (1546–1601), who was also one of the foremost scientists of his age. His observatory on the island of Hven became for about twenty years, from 1576, a mecca for astronomers, mathematicians, cosmographers and other scientists. As we saw in Chapter Three, among these was the Dutch cartographer, W. J. Blaeu, who later depicted the observatory and instruments on Hven on thirteen plates of his *Atlas Major*.

Scandinavian maps appeared in some of the early atlases, for example in the Ortelius *Theatrum Orbis Terrarum*, the Waghenaer *Spieghel der Zeevaert* (Plate 8), the *Caert Thresoor* of Barent Langenes (1598), the Mercator *Atlas* of 1595, and Blaeu's *Le flambeau de la navigation* (1620).

The height of Swedish cartography was reached by Anders Bure, who in 1626 published at Stockholm his map of Sweden, *Orbis arctoi nova et accurata delineatio*. It was engraved on copper and occupied six sheets. Seven copies are known. Bure had earlier (1611) published a map of the northern part of the country, but it was very simplified. Only two copies have survived. The 1626 map had a wide influence on the work of other cartographers, and was used by Visscher, Blaeu, Tavernier, De Wit and others.

After 1626 cartography in Scandinavia declined, although surveying was still practised. In 1628 Bure was ordered to organise a state survey of Sweden. The work continued until 1688, when the results were collated, and a new map produced, which was in many ways an improvement on that made by Bure alone. But it was incomplete, and Bure's original map had to be used to fill in the missing parts. The maps made in the survey were, moreover, looked upon as state secrets, so that access to them was difficult if not impossible. On the other hand, sea-charts based on the survey were published, and the Swedish soldier Graf Erik Jönsson Dahlberg (1625–1703) was permitted to produce an atlas from the surveys, but since only one copy was made (1698), any influence it might have had was curtailed. By 1735 conditions had eased and the official Surveyors' Office began to publish provincial maps, to be followed later by a map of the whole country.

John Mejer, a native of Holstein, and Geographer Royal to the King of Denmark made thirty-seven maps of Schleswig-Holstein, which were published in 1652 by the German publisher, Danckwerth.

Maps of the Scandinavian countries appeared in most later atlases including those of Speed (1627), Jansson (1629), Sanson (1666), De Wit (1680), De Fer (1700), and many others. Native cartographers apart from those already mentioned included Johann Månsson (*fl.* 1644), Peter Gedda (1661–97), Olasson Feterus (eighteenth century) and Ornehufud (eighteenth century).

☆   ☆   ☆

In ancient Russia, cartography came under the control of *Razryadny Prikaz*, an official department devoted to military matters. Tsar Ivan IV 'the Terrible' (1530–84) gave orders that the whole of Russia should be mapped, but there were no cartographers to do it; nor could foreign cartographers be persuaded to come to Russia. One map of Russia was, however, made in Italy by Paolo Giovio, on instructions from

Pope Clement VII.   In 1525 gifts from Grand Duke Vassily IV had been taken to the Pope by Vassily's ambassador, Demetrius Gerasimov (*circa* 1465–*circa* 1525), and Gerasimov gave Giovio the data on which to base his map.

Gerasimov was a colourful character.   He spent his childhood in Livonia, where he learnt German and Latin.   He entered the diplomatic service and served in Vienna, Prussia, Denmark and Sweden.   He also spent time at Rome and Florence, where he came into contact with leading scholars.   He was mainly interested in religion and helped in translations of a psalter and parts of the Bible; but his interests were wider than this, and he was much respected as a man of wide and versatile learning.

Another map of Russia, which was printed in 1546, has an interesting history which is worth giving in some detail.   Just before the time of Giovio's map, in 1516, the Emperor Maximilian I sent Sigismund Herberstein (1486–1566), a native of Styria, as his ambassador to Moscow.   Herberstein had studied at the University of Vienna and had obtained the degree of Bachelor of Laws.   After this he became an official, and in 1516 joined the diplomatic service, visiting many European countries including Switzerland, Denmark, Holland, Spain, Italy, Hungary, Germany, Poland and Russia.   He visited Moscow in 1517 and 1526.   While he was there, Herberstein got to know a boyar, Ivan Lyatsky, who later gave him an outline map of the country.   Lyatsky was a native of Lithuania who emigrated to Russia.   He took part in several Russian military campaigns, and in 1527 became Ambassador to Poland, and in 1528 Lord Lieutenant of Pskov.

Herberstein printed his map of Russia, which was probably based on that of Lyatsky in a version engraved by August Hirsvogel.   It was re-engraved several times. Lyatsky fell into disgrace and left Russia in 1534, settling at Vilna in Lithuania.   His map of Muscovy was engraved by Anton Wied in 1555.   As we have seen, he had already provided data to Herberstein, and his data was also used for a map published by Sebastian Münster in 1544.   A version was produced in 1570 by F. Hogenberg (Plate 50).

Early maps of Russia made in the west include that by Giacomo Gastaldi—*Moschovia nova Tabula*—in the 1548 edition of Ptolemy, published at Venice.   It was almost certainly based on the Gerasimov/Giovio map.   Another version appeared in the 1562 Latin edition of J. Moletius, which was published in Venice; this was reprinted in 1564 and 1574.

The first description of eastern Russia was made by the English traveller Anthony

Jenkinson, who was also the first Englishman to sail on the Caspian Sea and the first English ambassador to Russia. He was present on an expedition that sailed for Russia in 1557. It anchored on 12 July in that year in St. Nicholas road (anchorage), where Jenkinson landed and then made his way to Moscow. Some time later, on 23 April 1558, he set out, with letters of recommendation from the Tsar, for a journey towards Cathay. He was accompanied on this by two Englishmen, Richard and Robert Johnson, and a Tartar interpreter. Journeying partly by river, partly by sea, partly on land, he visited Nijni-Novgorod, now Gorky on the upper Volga; Astrakhan at the mouth of the Volga on the Caspian, and the limit of the Tsar's authority; the Mangishlak Peninsula, Sher Vezir, Urgendzh and Bokhara. On his return to England after this voyage, Jenkinson drew his map. Unfortunately no copy of it has survived, but copies were made for the atlases of Ortelius and De Jode. These are *Russiae, Moscoviae et Tartariae descriptio* (1562; Ortelius); *Moscoviae maximae amplissimique Ducatus chorographica descriptio* (1578; De Jode; Plate 52). There is also the map *Regionum Septentrionalium Moscoviam, Rutenos, Tartaros, eorumque hordus comprehendentium* (*circa* 1572; engraved by Johannes and Lucas à Deutecum), of which only one copy survives; it is at Leningrad. Its full title mentions Herberstein as a source in addition to Jenkinson.

Other maps of Russia engraved in Holland were made by Isaac Massa (based on Russian originals), and Hessel Gerritsz (1613; based on a manuscript map of the Tsarevitch Fyodor Godunov).[1]

There was in Amsterdam a Russian printing house, founded and run by a Dutch publisher, Jan Tessing, to whom Peter the Great granted a special licence to produce Russian maps in Holland. The first map printed there (it was also the first map printed in Russian) was published in 1699; it was Jacob Bruce's version of the Dutch Major-General George Mengden's map of southern Russia. Apart from the Russian version, one was also printed in Latin. It is a great rarity.

An atlas of the Don which was compiled by Cornelis Cruys, a Dutch admiral working for Russia, appeared in Holland in 1701. It was printed by Danckerts.

A general map of Muscovy, the 'Great Map' as it was called, was made in the reign of the Tsar Boris Godunov (1598–1605). There was only one copy which by 1627 had been worn out. A further 'Great Map' was made immediately, with a supplementary map of the Ukraine. To these were added a descriptive text and distance tables, the *Book of the Great Map*.

---

[1] *See* pages 41–2.

Although a Siberian map that was to accompany the second 'Great Map' was not made, the area was mapped in 1667 by Semyon Godunov of Tobolsk, at a time when Russia was beginning to expand eastwards. Revisions of his maps appeared in 1672 and 1687. Many further surveys of Siberia were made, and during the reign of Peter the Great (1689–1725) an atlas of the area was made and published by Semyon Ulanovitch Remezov.

Surveying made great advances under Peter's rule, and distant parts of the Empire, including the Kurile Islands, Kamchatka and the Caspian Sea, were mapped for the first time. The Caspian survey was carried out under the direction of the Russian sailor Karl von Verden; the resulting map was published in 1720 and was later copied by Ottens, Homann and others. Homann also published a map, on one sheet, of Kamchatka and the Chukchi Peninsula based on material given to him by the Russian surveyors Ivan Lvov and the Cossack Vladimir Atlasov.

Foreign craftsmen, such as Peter Picard (1670–1737) and Adriaen Schoonbeck (d. 1705), were employed in the eighteenth century to reproduce original maps by native cartographers and others. Russian engravers were also beginning to work on maps. Among them was Alexei Zubov (1682–1743), a pupil of Schoonbeck, who engraved several Russian maps between 1705 and 1736. There was also Vassily Kiprianov (d. 1723) of Moscow, engraver, cartographer and the first private publisher of maps in Russia, whose engraved maps were made between 1706 and 1717.

An important survey of Russia was undertaken between 1717 and 1734 by Ivan Kirilov (1689–1737), who had in 1717 been put in charge of all Russian cartography, as Director of the Russian Cartographic Office. From 1702 to 1707 Kirilov attended a naval school founded in 1701 by Peter the Great for the training of young men in mathematics and navigation. From 1707 he served in the Navy Department, during which time he was sent abroad to widen his knowledge, spending part of the time at Amsterdam and London. Later he transferred to the Civil Service and served in the Land Survey Department.

Kirilov needed expert advice in connexion with his survey, and with this in view invited Joseph Nicolas de l'Isle (1688–1768), brother of Guillaume de l'Isle (see Chapter Four) to come to Russia. The two men were at variance on how best to achieve their aim. De l'Isle wanted to carry out an astronomical survey first, to obtain a number of astronomically-determined points on which to base his geographical survey. Kirilov thought this would be too painstaking to achieve quick results. He therefore followed his own method of making a survey along roads and rivers with measurement of distances, all to be later checked by astronomically-obtained positions, with de l'Isle as an adviser. The result of Kirilov's labours was the first atlas of all Russia. Its first part, a general map and fourteen regional maps, appeared in 1734 under the title of *Atlas Imperii Russici* . . . and was dedicated to the Empress Anna Ioannovna. Its general map was the earliest one to give a reasonably accurate picture of Russia, including its rivers and sea coast. The complete atlas should have been in three volumes, each containing one hundred and twenty maps, but when Kirilov died of consumption in 1737, work on it was abandoned, and Kirilov's widow sold his records and plates to the Academy of Sciences. Today only one copy of the first issue survives; it is preserved in Moscow. De l'Isle's copies of the maps are in the Bibliothèque Nationale, Paris.

Attempts were made after Kirilov's death to have the atlas published, but de l'Isle was not sufficiently satisfied with its presentation to allow this. But during de l'Isle's absence from Moscow, the Academy of Sciences rushed an edition through the press, and it finally appeared, under his name, in 1745. It contained a general map and nineteen regional maps. Like most maps of the time, those in the Russian atlas were imperfect, but corrections were made between 1757 and 1765 by Michael Lomonsov (1711–65), the Director of the Geographical Department of the St. Petersburg Academy of Sciences. Subsequent maps of Russia produced by western cartographers were based on these maps.

BIBLIOGRAPHY

BAGROW, LEO: 'At the source of the cartography of Russia'. *Imago Mundi*, Vol. XVI, pp. 33–48. 1962.

—— 'The first map printed in Russian'. *Imago Mundi*, Vol. XII, pp. 152–6. 1955.

—— *History of Cartography*. 1964.

—— 'Ivan Kirilov, compiler of the first Russian atlas, 1689–1737'. *Imago Mundi*, Vol. II, pp. 78–82. 1937.

—— 'Semyon Remezov—a Siberian cartographer'. *Imago Mundi*, Vol. XI, pp. 111–26. 1954.

FLOROVSKY, A.: 'Maps of the Siberian route of the Belgian Jesuit, A. Thomas (1690)'. *Imago Mundi*, Vol. VIII, pp. 103–8. 1951.

KEUNING, J.: 'Jenkinson's map of Russia'. *Imago Mundi*, Vol. XIII, pp. 172–5. 1956.

LYNAM, E.: 'The early maps of Scandinavia'. *Geographical Journal.* 1927.

—— *Early maps of Scandinavia and Iceland.* 1934.

—— *The Carta Marina of Olaus Magnus Venice 1539 and Rome 1572.* 1949.

SPEKKE, A.: 'A brief cartographic/cosmographic view of the eastern Baltic coast up to the 16th century'. *Imago Mundi*, Vol. V. 1948.

TOOLEY, R. V.: *Maps and map-makers.* 1962.

CHAPTER SEVEN

Africa

THAT part of Africa which borders the Mediterranean, from Egypt in the east to Mauretania in the west, has played a formidable part in the history of the world since remote pre-christian times. Its southernmost tip, in the area of the Cape of Good Hope, was not discovered until 1487, when Bartholomew Diaz sailed round it. The interior was not known until much later. Indeed, during a lecture delivered in 1925 to the British Association, the geographer A. R. Hinks mentioned that Natal had still not been properly mapped.

We have already dealt with some of the early exploration of the coasts of Africa, but we may here concentrate on it in greater detail. The history of this exploration has been distorted because we have too little in the way of surviving evidence on which to base it. Out of the great Portuguese work in this direction, for instance, only two complete fifteenth-century charts remain. Apart from these, records of Portuguese discoveries are depicted only on Italian maps, the creators of which had drawn on information from Portuguese sources. These Italian maps (they were mostly drawn by Venetian cartographers) include the world map made in 1459 by Fra Mauro (d. 1460), which was completed with the assistance of Andrea Bianco (*fl.* 1436–59), a Venetian galley-master who traded with the Low Countries and had in 1448 made a chart showing the coasts of Africa down to Cape Verde and Senegal. Mauro's map shows also an open seaway round the south coast of Africa, and gives East Africa in some detail. It is thought that his information here was derived in part from Arab sources.

Charts and atlases recording discoveries made on voyages directed by the Portuguese Prince Henry the Navigator (1394–1460), Duke of Viseu, Governor of Algarve, and 'protector of Portuguese studies', were made by Grazioso Benincasa (*fl.* 1450–82) of Ancona. These discoveries included Sierra Leone. Details of the Gold Coast were recorded about 1471 by João de Santarem and Pedro de Escobar. The earliest existing signed Portuguese chart, by Pedro Reinel (*fl.* 1485–1535), records details gathered on Diogo Cão's voyage to the Congo in 1482–4 (Plate 3; see also page 26).

Because of its long and fragmented history, the mapping of Africa has been uneven. The maps of its northern parts were quite accurate portrayals in classical and mediaeval times, but, when any attempt was made to show the more remote parts, guesswork was often resorted to. For instance the Indian Ocean was for long shown as land-locked, Africa being shown joined to an imaginary southern continent (Plate 1). Sometimes the southern parts were not shown at all, as in all editions of Ptolemy before that of 1508. In the Strasbourg edition of 1513 the southern area, complete with the Cape, is shown. But the Cape had been shown earlier than this, on Martin Behaim's globe of 1492, and as early as 1306 a portolan chart attempted to show that the South Atlantic and Indian Oceans were confluent.

Early printed maps of the whole African continent were made by Contarini (1506), Waldseemüller (1508), and George Reisch (in the 1508 Ptolemy world map). Others appeared in geographical textbooks such as *Globus Mundi* (Strasbourg, 1509), *Itinerarium Portugualesium* (Milan, 1508; in this case it is part of the title page), and Reisch's *Margarita Philosophica* (1513). Especially popular with collectors, for it is normally the earliest available map of Africa, is the wood-cut map published in Sebastian Münster's *Cosmographia* (Plate 51). It is a picturesque production and shows an elephant, exotic birds, and a great cyclops. It could be taken as an illustration of Jonathan Swift's lines:

> So geographers, in Afric-maps,
> With savage-pictures fill their gaps;
> And o'er unhabitable downs
> Place elephants for want of towns.

Between 1560 and 1600 there was a considerable increase in the number of published maps of Africa, such as those by Gastaldi (1560, 1564), Ortelius (1570), De Jode (1593), Mercator (1595), Linschóten (1595), Quad. Bussemacher (1600). Some of these were reprinted many times, and nearly all have great decorative value.

Even more maps of Africa were produced in the following century, including those of Speed (1626–7) which was the first English map of Africa, Blaeu (1630, 1650), Hondius (1631, 1659), Jansson (1632, 1650), Sanson (1650, 1656, 1669, 1685), Cluver (1659), Danckerts (1660), De Wit (1670, 1680), Ogilby (1670), Allard (1680), Moll (1700) and de l'Isle (1700).

Some of these, including those by Blaeu, Hondius, Jansson and Sanson, went into many editions, and many of them were highly decorative. Blaeu's map of Africa from his *Grooten Atlas*, for example, has a border of natives in colourful costumes, and bird's-eye views of prominent cities. Elephants, sheep and other inhabitants are shown on land, and the oceans have their ships and monsters, including an elegant hippo-campus off Cape Negro on the west coast (Plate 12). Marine charts of the oceans around Africa were also issued by such publishers as Seller and Blaeu.

The eighteenth century saw the publication of a large number of maps of Africa, including those of Moll (1714), de l'Isle (1722), Homann (1730), Bowen (1744, 1760), D'Anville (1749), Vaugondy (1756), and Kitchin (1794). On the whole they are less decorative than earlier maps, but their cartouches and scales are often more elaborate and show large figures in native dress, exotic animals such as crocodiles and elephants, or inset views. As the century wore on the maps showed more details of the interior of the continent as these were discovered by explorers.

As in the case of nineteenth-century maps of Europe, maps of Africa are usually plainer than their predecessors, with little or no decoration. Yet they also achieve a certain decorative quality from the excellence of their engraving. Outstanding maps of the period are those by Adrien Hubert Brué (1786–1832; map published 1800), Aaron Arrowsmith (1802, 1811), Pierre Lapié (1777–1851; map published 1803), F. L. Güssefeld (1804), and Cary (1805, 1811, 1828).

Regional maps of the continent were issued from early times. For instance, maps of North Africa were published by Münster (1540), Blaeu (1640), Homann (1730) and Reinicke (1802); of South Africa by Linschóten (1598), Blaeu (1640), de l'Isle (1720), Reinecke (1802), Tallis (1840) and Arrowsmith (1842); of West Africa by Langren (1596), Blaeu (1640), Ogilby (1670), Homann (1730), Jefferys (1768), and Tallis (1840); and of East Africa by Ortelius (1572), Blaeu (1640), Bowen (1747) and Tallis (1850). These are but a few examples from dozens that exist. There are also maps, by many of the great cartographers, of the African islands—the Canaries, Madagascar, the Azores, Cape Verde and St. Helena.

BIBLIOGRAPHY

LANE-POOLE, E. H.: *The Discovery of Africa . . . as reflected in the maps . . . of the Rhodes-Livingstone Museum.* 1950.

LYNAM, E.: 'The Discovery of Africa (Mon. Carto. Africae et Aegypti)'. *British Museum Quarterly.* 1927.

MAJOR, R. H.: 'On the map of Africa published in Pigafetta's Kingdom of Congo, 1591'. *Royal Geographical Society Proceedings.* 1867.

TOOLEY, R. V.: 'Collectors' Guide to the Maps of Africa. Parts I and II'. *Map Collectors' Series*, Vol. V.

—— 'Early Maps and Views of the Cape of Good Hope'. *Map Collectors' Series*, Vol. I.

—— *Maps and map-makers.* 1962.

—— 'Printed Maps of Africa. Part I, The Continent of Africa, 1500–1600. Part II, Regional Maps, 1500–1600'. *Map Collectors' Series*, Vol. III.

CHAPTER EIGHT

Asia

O F all the countries of Asia that with the oldest civilisation and the longest tradition of cartography is China. The Chinese knew many of the techniques of surveying before they were known in the West—for example the astronomical determination of places, the principles of the compass and of the gnomon, and the technique of levelling. There were, however, some considerable gaps in general geographic knowledge. The early Chinese thought that the world was square, although by about 400 B.C. its true shape began to be realised.

A map of the whole kingdom of China was made before 1100 B.C., in addition to many smaller regional and specialised maps. So important at about this time were the activities of cartographers deemed to be that special regulations were drawn up to control them. Afterwards cartography declined in China for a time, but by about 450 B.C. quality had returned, and an official topographical description of the country, with maps, was issued.

During the second half of the third century B.C., at the time of the Ch'in Dynasty, the provincial borders of the country were redrawn, which necessitated new surveys and revisions. At this time, too, maps became first-class works of art. Some were made of bamboo, others painted on silk. Another great advance came with the invention of paper at the end of the second century B.C. Thereafter, cartography became the responsibility of a government department which some time later issued a cartographer's manual by P'ei Hsiu. P'ei Hsiu (A.D. 224–271), known as 'the father of Chinese cartography', was the first great Chinese mapmaker. He occupies a position in Chinese cartography not unlike that of Ptolemy in the West. His works included a map of China on eighteen sheets to a scale of 500 li[1] to one inch. No copy survives.

---

[1] I li=approximately ⅓ mile.

In fact the earliest extant Chinese maps are engraved on stone and date from the twelfth century A.D.

In the centuries following P'ei Hsiu, Chinese cartography was influenced by the introduction from India of Buddhism. For instance in a number of maps the world was shown in the form of a disc surrounded by ocean, as in some early western maps. Another influence was the expansionist policy of the T'ang Dynasty, which resulted in the compilation of a geographical description with maps, in sixty volumes, of the lands in the West.

The development of cartography in China was arrested during the Mongol invasions of the early thirteenth century, though some developments took place later, in the later thirteenth and fourteenth centuries under the Mongol Dynasty, during which period the craft was much influenced by astronomy. At this time an atlas of the provinces of China was compiled by Chu Ssu-pên (*fl.* 1311–12). This in turn had an influence on western cartography, for it was used as a model for the *Atlas Sinensis* of the Jesuit Martino Martini (1614–61), published by Joan Blaeu, Amsterdam, in 1655. This held its place as a standard in European maps of China until D'Anville's *Nouvel Atlas de la Chine* appeared in 1737 (*see below*).

Jesuit missionaries had a considerable influence on Chinese cartography in the sixteenth century. Among them was Father Matteo Ricci (1552–1610), who made a Chinese language version of a European world map, probably based on Ortelius and Gastaldi. (De Jode's copy of Gastaldi's 1561 map of East Asia, which he made in 1578, is shown in Plate 7.) A unique impression of such a map is in the Biblioteca Ambrosiana in Milan; it may be Ricci's actual map, which was reissued in a second edition in Nanking in 1599. Also made by Ricci, and issued in Peking in 1602, was a new world map in six strips. It was printed from woodblocks and measured twelve by six feet. A later version of this, painted on silk, is preserved at Peking.

Sea maps and charts formed an important part of Chinese cartography, and some of them took in details of oceans and coasts far from China itself. One example, made in the sixteenth century, depicted the coasts from Amoy to the Persian Gulf. Another, made in the eighteenth century, showed the coasts from Amoy to Korea.

There are two main types of Chinese sea-map: the ordinary type, orientated to the north, and a narrow strip type showing a sector of the coast running horizontally from right to left, irrespective of the points of the compass. Specimens of the latter are known from as early as about 1422, and as late as 1884. They were used for navigation, for military and naval purposes, and as aids in administration.

Another big advance was made in Chinese cartography during the period of the Manchu Dynasty, which began in the middle of the seventeenth century. A new map of the country was made by order of the Ministry of War about 1655, and it was published in various versions until 1900. Jesuit missionaries were called upon to help during the whole of the seventeenth century, particularly during the reign of the Emperor K'ang-hsi, who ascended the throne in 1671. Among them was Ferdinand Verbiest (1623–88) whom K'ang-hsi appointed *Summus Praefectus Academiae Astronomicae*, a post he held from 1671 until 1685. During this time Verbiest made a world map in hemispheres, engraved on wood, in six rolls; it was printed at Peking. It was taken from Chinese sources for China itself, and from European sources for the rest of the world. Later K'ang-hsi commissioned a detailed map of China from the Jesuits, which was made during the first quarter of the eighteenth century. Part of the source material upon which this map was based consisted of two hundred and thirty-one earlier regional Chinese maps. Copies of these were taken, various additions made, and the final map appeared about 1717 or 1718 in over thirty wood-cut sheets. Some of these were sent to Europe, and became the basis for d'Anville's *Nouvel Atlas de la Chine* (forty-two maps) which was published at The Hague in 1737 (Plate 54).[1] It contained maps of countries adjacent to China in addition to those of China itself, as may be gathered from its full title *Nouvel Atlas de la Chine, de la Tartarie Chinoise, et du Thibet*. Some of the material it contained remained standard until the present century.

So far as European ideas of China are concerned, the first important information of its shape and size was brought back in the thirteenth century by the traveller Marco Polo. The idea of its shape became more definite in the sixteenth century as a result of information brought back by Jesuit missionaries and Portuguese traders.

The first map of China produced in Europe appeared in 1584. It was compiled by the Portuguese Jesuit Ludovico Georgio (1564?–1613), and issued by Ortelius in his *Theatrum Orbis Terrarum Additamentum III*. This map was used as a general source of information long after this by such cartographers as Hondius, Mercator, de Jode and Linschóten. Needless to say, all of these maps, and others, contained inaccuracies, for most of the material on which the first of them was based was secondhand at best.

Size was one great difficulty. In some maps, China is shown three, four, or even five times as large as Europe; on others it is of about the same size. Moreover, on some of them China and Cathay were treated as separate countries, whereas it had been proved by the Jesuits that they were the same. Nevertheless, a more accurate represen-

[1] *See also* page 49.

tation was gradually achieved. Hondius, for instance, in his general map of Asia of 1631, shows Korea as a peninsula, whereas it had hitherto been shown as an island of varying shape. The large curve of the coast of China was shown for the first time in 1596 by Linschóten.

Other important European maps of China were issued by Joannes van Loon in the seventeenth century and by J. B. Du Halde in his *Description de la Chine* (1735; four volumes), which contained views in addition to maps. Another map, drawn by Nicolas d'Abbeville Sanson, was published, in *Abbrégé De La Carte De La Chine*, by Pierre Mariette of Paris in 1670, three years after Sanson's death. It was based on an earlier map, *Mappa Imperii Sinarum* by a Polish missionary Michael Boym (1612–59). Boym was the first European cartographer to insert Chinese characters with their nearest equivalent in roman characters. They appear thus on Sanson's map.

Sanson drew other maps of China based on maps by Ricci and Michele Ruggieri (1543–1607). He also printed one by Samuel Purchas (1575–1626) in *L'Asie* (1652, 1658 and 1683); this was taken from the original in *Purchas His Pilgrimes* (four vols., London, 1625), but considerably changed.

Martini's 1655 map of China, to which reference was made earlier in this chapter, reappeared in Richard Blome's *A Geographical Description of the Four Parts of the World taken from the Notes and Workes of the Famous Monsieur Sanson* (London 1670). A map of China appeared, too, in *The History of That Great and Renowned Monarchy of China* (London 1655), a translation of an Italian work. The map is entitled 'An Exact Mapp of China, being faithfully Copied from one brought from Peking by a Father Lately resident in that City. 1655'. It was drawn by Alvarez de Semedo (1585–1658), author of the original Italian work from which the translation was made.

☆   ☆   ☆

Japanese cartography originated later than Chinese—that is during the seventh and eighth centuries A.D.—and developed independently of it. The first general map of Japan is said to have been made by Gyogi-Bosatsu (670–749). It is known today by a reproduction published in the 'Shugaisho' Encyclopedia (1596–1614). Gyogi-Bosatsu was a Korean Buddhist missionary who came to Japan early in his life, and taught the people the arts of road and canal construction, and of bridge and dam building. His map was adopted as a standard pattern, known as the Gyogi type, and was in use for eight centuries. It influenced Korean, Chinese and European maps, as well as those produced in Japan.

The earliest surviving Japanese map is much later than Gyogi-Bosatsu's. It is a fourteenth-century copy of one belonging to the late eighth or early ninth centuries. It is preserved in a temple (Ninni ji) near Kyōtō. About two hundred years later, after this copy was made, Japanese cartographers began to get sight of European maps. These had a great influence on Japanese cartography during the early part of the seventeenth century, when a number of world maps were made in Japan, based on European types—probably on maps by Gastaldi and Ortelius, and brought in by Christian missionaries. Sometimes they were used for the decoration of screens. There were other various types of Japanese world maps; some were based on Buddhist cosmology, some were influenced by Chinese thought, others showed Japan as the centre of the world, and others illustrated grotesque legends.

From 1643 the Japanese government followed a strict isolationist policy, nationals being forbidden on pain of death to leave the country at all. To make this ban even more effective, the building of large ships was forbidden. Not only did this make improved sea-charts superfluous, but it put a stop to any serious Japanese development of the cartography of foreign parts. It did on the other hand enable the Japanese to concentrate more effectively on maps of the homeland, many of which reached a high state of artistic excellence. Contact with Europe was not resumed until the latter part of the eighteenth century.

Ishikawa Toshiyuki (Ryūsen; *fl.* 1688–1713) was responsible for some really beautiful maps and town plans, though cartographically they are inaccurate. Better in this respect are those of Nagakubo Genshu (Sekisui), whose map of the whole of Japan is particularly notable and who was the first Japanese to use scales, parallels and meridians. Apart from this his maps show little European influence. Certain other Japanese cartographers did come under European influence, in particular Shiba Kokan, publisher in 1792 of the first Japanese copper-engraved map.

Topographical surveys of Japan were first made in 1605, and others were made until 1843. The science of surveying was highly developed by the Japanese, the methods followed being comparable with those used in Europe, even to the instruments used. None of the maps made during these surveys was available to the public, and one surveyor, Inō Chūkei (1745–1818), a rich brewer, prepared well over two hundred regional maps on his own initiative between 1800 and 1816. His maps have remained in use until the present day.

Japanese printed maps were first made in the seventeenth century. The oldest surviving specimen of a printed general map of Japan—it is of the Gyogi type—dates

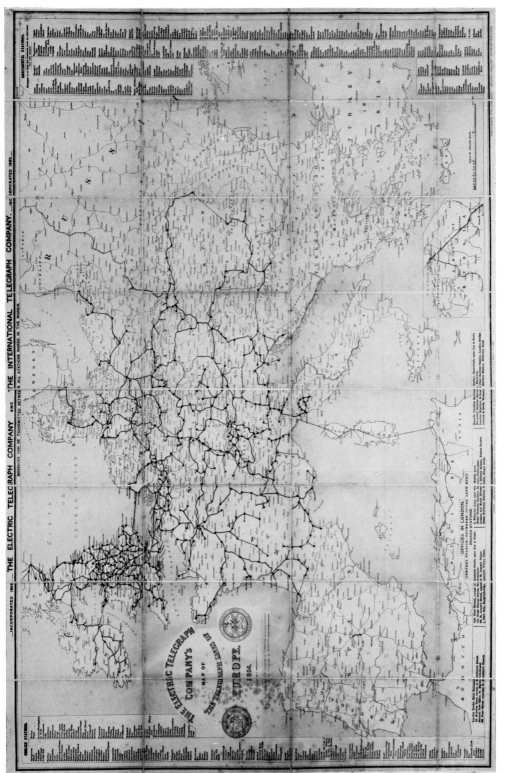

53. The Electric Telegraph Company's Map of The Telegraph Lines of Europe, 1854. Lithographed by Day & Son.
Private Collection.

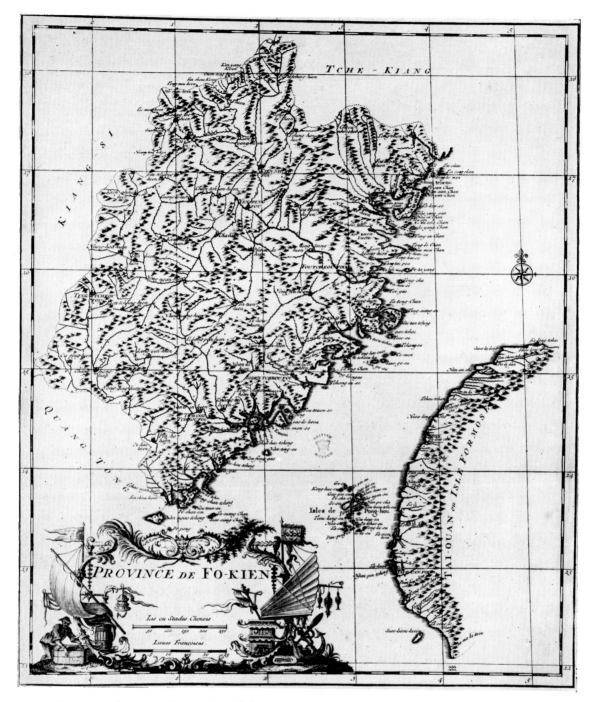

54.   Map of Fo-Kien from D'Anville's *Nouvel Atlas de la Chine*, 1737.   *British Museum.*

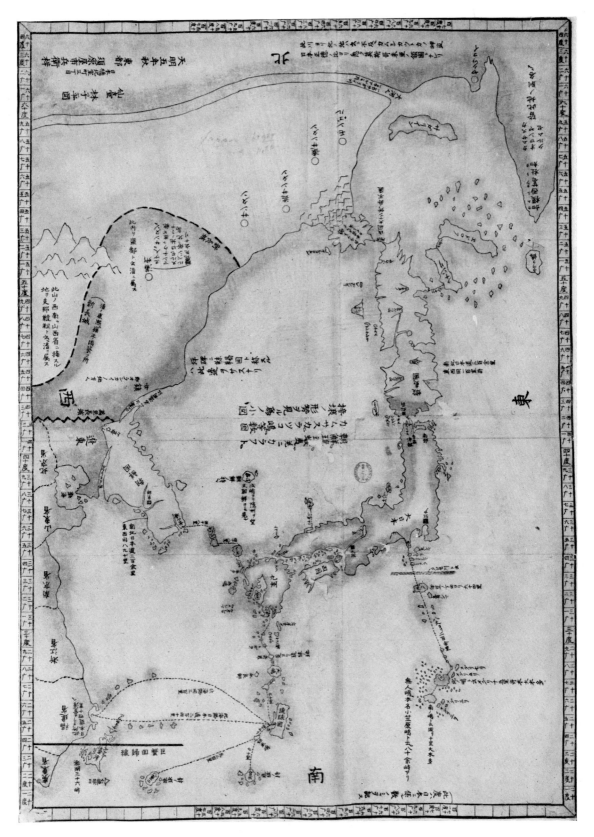

55.   Map of Japan and the surrounding islands, showing the various sea routes.
Drawn by Hayashi Kohei, 1785.   *British Museum.*

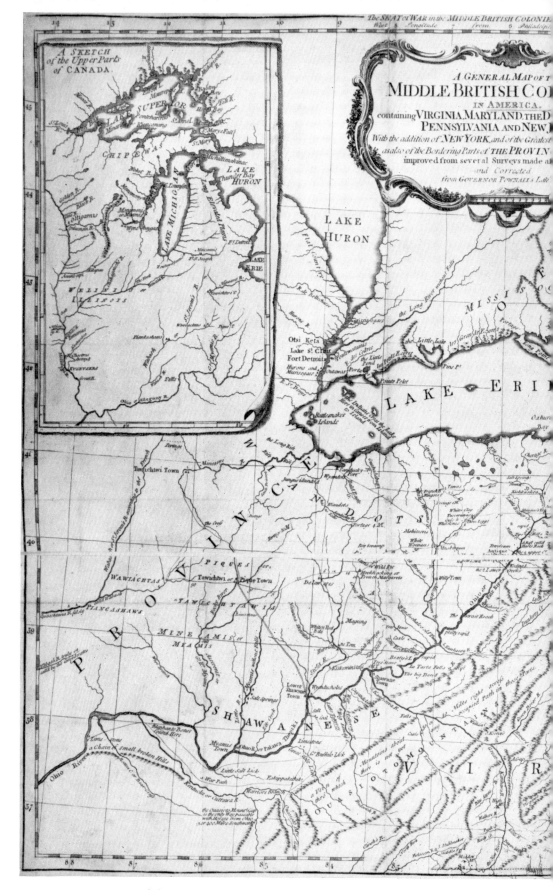

56. Map of the Middle British Colonies in America by Lewis Evans, 1756.

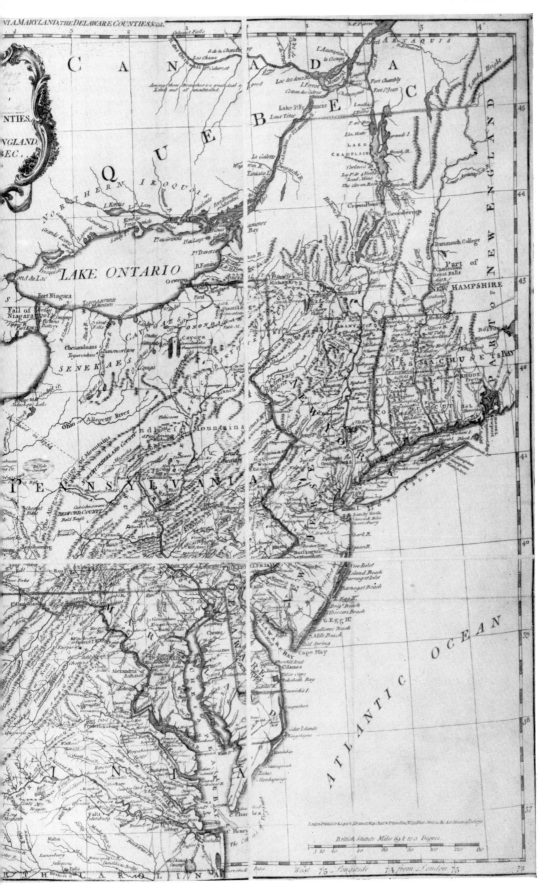

British Museum.

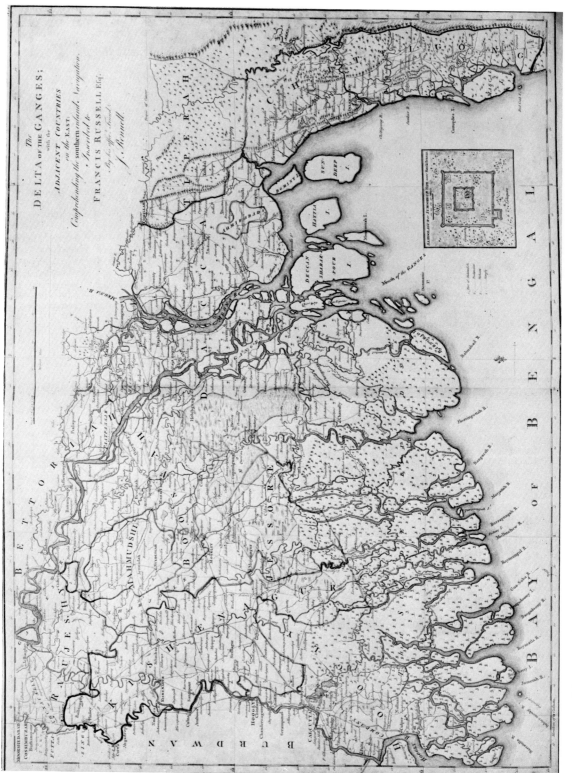

57.   Map of the Delta of the Ganges from Rennell's *Bengal Atlas*, 1781.

British Museum.

58. Woodcut map of The New World from Sebastian Münster's edition of Ptolemy, 1545. *British Museum.*

from 1651. It is in the Imperial University of Kyōtō. Printed maps were mostly made from woodcuts, and are closely related to the *Ukiyō-ye* prints (prints of 'the floating world': the demi-monde). Some of the great *ukiyō-ye* artists, like Hokusai, produced maps (Plate 55). It is a fascinating branch of the subject and one that should attract a collector who likes unusual material. Hokusai (1760–1849), 'the old man mad about painting', made, in his eighty-first year, a poetic map of China, drawn in perspective. It was published, probably at Edō (Tōkyō), by Seiundo and engraved by Egawa Sentaro. Hishikawa Moronobu (*circa* 1618–94) designed a map of the Tōkaidō Highway in 1690. Road maps were made, too, by Utagawa Hiroshige (1797–1858), and mountaineers' maps were made by the nineteenth-century artist Sadahide.

During the nineteenth century, maps on porcelain plates were sometimes made in Japan. They are very attractive, but rare. Though cartographically distorted, they contain a large amount of detail. Such maps were obviously not made for practical purposes, but for their artistic interest.

In European maps Japan first appeared on Fra Mauro's fifteenth-century *mappa mundi*,[1] but it was a long time before its true shape was shown even approximately. It was shown by Ortelius in 1570, on his map of the East Indies, as one large island with an archipelago to north and south, and, on his map of Asia of the same date, in an entirely different shape. Other cartographers were equally wide of the mark. Somewhat nearer reality was Ortelius's separate map of Japan, compiled by the Portuguese Father Ludovico Texeira (1595), based on the Gyogi type. It was accepted as the standard for a long time. Mercator re-engraved it, a version that was appropriated and re-issued by Jansson about 1650. A new map was issued about 1700 by Schenk and Valk, and in 1750 Sieur Robert issued a map that showed the Japanese archipelago in its more or less true shape. Accurate maps of parts of it—the Ryukyu Islands and Southern Japan—by the Jesuit Antoine Gaubil (1689–1759) were published in *Lettres édifiantes et curieuses* (Paris 1758). Further maps of the area were published in 1818 in Captain Basil Hall's *Account of a Voyage of Discovery to the West Coast of Korea and the Great Loo-Choo Island.*

☆   ☆   ☆

Of native Indian cartography there is little that one can say beyond the fact that it is negligible. What there was of it was purely symbolic and was based on the Buddhist view of cosmography. On the other hand, European maps of India are numerous.

[1] *See* pages 21–2.

They appear in most atlases from the time of Ptolemy, although the shape of the sub-continent was at first considerably distorted.   The first approximately correct representation was made in a world map by Reisch (1508).   Other early maps of the country were made by Mercator, Ortelius, Gastaldi (Plate 7), and Bertelli.   Dutch maps of India include those by Blaeu, De Wit, Hondius, Jansson, Van Keulen and Visscher. The Germans Homann and Seutter, and the Frenchmen Bellin, D'Anville, de l'Isle, Jaillot and Sanson also issued them.

The English naturally have had a special interest in Indian maps, and it was by English cartographers that the greatest number was issued.   The first one was drawn by William Baffin, engraved by Renold Elstrack and published in 1619 by T. Sterne.   It was copied by others on at least two occasions.   Later English maps of India were issued by Bowen, Moll and many others.   Outstanding were those of James Rennell (1742–1830), who became Surveyor-General of Bengal.   He made the famous *Bengal Atlas* (1779) of twenty-one maps and plans (Plate 57).   His other works include *An Actual Survey of the Provinces of Bengal, Behar* . . . (1776, 1794), a *Map of Bengal, Behar, Oude, Allahabad* . . . (1776, 1786, 1794 and 1824), a *Map of Hindostan* (1782, 1794) and a *Map of the Provinces of Delhi, Agrah, Oude and Allahabad* (1794).   Aaron Arrowsmith issued a map of India in six sheets (1804) and an *Atlas of South India* (eighteen sheets; 1822).   There were many more such maps.

BIBLIOGRAPHY

Auysawa, Shintaro: 'The types of world maps made in Japan's Age of National Isolation'. *Imago Mundi*, Vol. X, pp. 123–7.   1953.

Baddeley, J. F.: 'Father Matteo Ricci's Chinese World Maps, 1584–1608'.   *Geographical Journal*, Vol. L.   1917.

Bagrow, Leo: *History of Cartography*.   1964.

Baker, J. N. L.: 'Some original maps of the East India Co'.   *Konin.   Nederl.   Aardrijkskundig Genootschap 2 Serie deel LIII.*   1936.

Beans, George H.: 'An Early Plan of Kyōtō'.   'The Grid Pattern in Old Chinese Maps'. 'The Orientation of Japanese Maps'.   *Imago Mundi*, Vol. XI, p. 146.   1954.

—— 'Japan is a Mountainous Country'.   *Imago Mundi*, Vol. IX, p. 108.   1952.

Campbell, Tony: 'Japan: European Printed Maps to 1800'.   *Map Collectors' Series*, Vol. IV. *Catalogue of Maps, Plans etc. of India and Burma and other parts of Asia.*   1891.

CRESSEY, G. B.: 'Evolution of Chinese Cartography'. *Geographical Review*. 1934.

CRONE, G. R.: 'Seventeenth Century Dutch Charts of the East Indies'. *Geographical Journal*. 1943.

GERINI, G. E.: *Researches on Ptolemy's Geography of Eastern Asia*. 1909.

HEAWOOD, E.: 'The Relationships of the Ricci Maps'. *Geographical Journal*, Vol. L. 1917.

IKEDA, TETURO: 'World Maps in Japan before 1853'. *Comptes Rendus: International Geographical Congress Amsterdam*. 1938.

KISH, GEORGE: 'Some Aspects of the Missionary Cartography of Japan during the 16th Century'. *Imago Mundi*, Vol. VI, pp. 39–47. 1949.

—— 'The Japan on the "Mural Atlas" of the Palazzo Vecchio, Florence'. *Imago Mundi*, Vol. VIII, pp. 52–4. 1951.

KITAGAWA, KAY: 'The Map of Hokkaido of G. de Angelis, *c.* 1621'. *Imago Mundi*, Vol. VII. 1950.

MANCHESTER, CURTIS A.: 'A Tokugawa Map of Japan on Porcelain'. *Imago Mundi*, Vol. XVI, pp. 149–51. 1962.

MCGOVERN, MELVIN P.: 'A List of Nagasaki Maps Printed during the Tokugawa Era'. *Imago Mundi*, Vol. XV, pp. 105–10. 1960.

MILLS, J. V.: 'Chinese coastal maps'. *Imago Mundi*, Vol. XI, pp. 151–68. 1954.

MUROGA, NOBUO and KAZUTAKA, UNNO: 'The Buddhist World Map in Japan and its contact with European maps'. *Imago Mundi*, Vol. XVI, pp. 49–69. 1962.

NAKAMURA, HIROSI: 'Old Chinese World Maps preserved by the Koreans'. *Imago Mundi*, Vol. IV, pp. 3–22. 1947.

—— 'The Japanese Portolanos of Portuguese origin of the XVIth and XVIIth century'. *Imago Mundi*, Vol. XVIII, pp. 24–44. 1964.

NEEDHAM, J.: *Science and civilisation in China*. Vol. 3. *Mathematics and the sciences of heaven and earth*. 1959.

RAMMING, M.: 'The Evolution of Cartography in Japan'. *Imago Mundi*, Vol. II, pp. 17–21. 1937.

SOOTHILL, W. E.: 'Two Oldest Maps of China extant'. *Geographical Journal*, Vol. LXIX. 1927.

SZCEŚNIAK, BOLESLAW: 'Matteo Ricci's maps of China'. *Imago Mundi*, Vol. XI, pp. 127–36. 1954.

—— 'The Antoine Gaubil maps of the Ryukyu Islands and Southern Japan'. *Imago Mundi*, Vol. XII, pp. 141–9. 1955.

—— 'The Seventeenth Century Maps of China', *Imago Mundi*, Vol. XIII. pp. 116–36. 1956.

TOOLEY, R. V.: *Maps and map-makers*. 1962.

UHDEN, R.: 'The Oldest Portuguese Chart of the Indian Ocean, 1509'. *Imago Mundi*, Vol. III. 1939.

CHAPTER NINE

America

DETAILED mapping of America, as distinct from the general mapping of its coastlines, was begun under the colonial administrations of the sixteenth and seventeenth centuries. The Spaniards in particular found it necessary to have fully detailed maps because of their bureaucratic control, which reached from the King, the Council of the Indies and the Board of Overseas Trade in the homeland, to the Viceroyalties of Peru and Mexico in the New World. In this way their possessions which stretched from Chile in the south to California in the north-west and Florida in the north-east, were apportioned and administered. Brazil was administered by Portugal.

All of these vast regions were mapped in varying degrees of detail in the sixteenth and seventeenth centuries. So too were the tracts of country administered by the English, French and Dutch on the Atlantic coast, in the Gulf of St. Lawrence, and inland from the Mississippi up to Hudson Bay. The most accurate sixteenth-century map of a part of North America was made between 1585 and 1587 by the Englishman John White, when he drew the coast from Chesapeake Bay in Maryland and Virginia, down as far as Florida. A map of the area north from Chesapeake Bay up to the Great Lakes, Nova Scotia and the Gulf of St. Lawrence, was made about a quarter of a century later than this (1613) by Samuel de Champlain, a French geographer. A further map of the area was made by de Champlain in 1632.

The first map actually made in America was a woodcut map of New England by John Foster, a printer of Boston. It was issued in 1677. A well-known woodcut map of a later period was that of the borderland of Maryland and Pennsylvania. It was made in 1733 by Benjamin Franklin. Woodcut American maps are not numerous, the majority of early ones being engraved on copper. The first of such maps to be produced in America was by Captain Cyprian Southack, and was issued at Boston in

1717. It was his *New Chart of the British Empire in North America*. Some claim that the highest point of colonial cartography was reached in 1755 when a map by Lewis Evans (*circa* 1700–56) was issued, entitled *Map of the Middle British colonies in America* (Plate 56). It was engraved on copper by James Turner and printed on a press owned by Benjamin Franklin. Impressions on silk are known.

Little accurate surveying of the interior of North America took place before the middle of the eighteenth century. There then developed a demand for more reliable maps, and surveyors were encouraged by the Lords Commissioners for Trade and Plantations to produce them. Lewis Evans's map was among those which resulted from this. So also was his *Map of Pensilvania, New-Jersey, New-York . . .* (1749).

Among later American cartographers were the prolific Careys of Philadelphia: Mathew (1760–1839) and Henry Charles (1793–1879) and their associates Isaac Lea (1792–1886) and Joseph C. Hart (d. 1855). Their works include *Carey's American Atlas* (1796, 1801, 1805, 1813, 1814), *Carey's General Atlas* (1796, 1802, 1814, 1817, 1818, etc.), *A Complete historical, chronological and geographical American Atlas* (1822, 1823, 1827), *A Scripture Atlas* (1817), *A General Atlas for the present War* (1794) and *Family Cabinet Atlas* (1832 and 1834). Included among engravers and cartographers who worked for the Careys were J. Finlayson, F. Lucas, S. H. Long, E. Paguenard, W. Barker, Amos Doolittle, Harding Harris and General D. Smith. In many ways the Careys may be compared with their near-namesakes in England, the Carys (*see* pp. 79–80), for like that family, they provided the public with a wide choice of maps and atlases produced by a wide-ranging team.

Thomas Jefferson (1743–1826), the third President of the U.S.A., produced a map of four of the states of the eastern seaboard in 1786. It is said to be the best surviving eighteenth-century map of the area. It was engraved partly in London by Samuel Neele (1758–1824), and partly in Paris by Guillaume Delahaye (1725–1802), who corrected some initial errors. It is entitled *A map of the country between Albemarle Sound, and Lake Erie, comprehending the whole of Virginia, Maryland, Delaware and Pennsylvania*. Henry Schenk Tanner (1786–1858) of Philadelphia, and his associates, Tanner and Marshall; and Tanner, Vallance, Kearney and Co. were an important group of American cartographers. Their products include *Atlas classica* (1840), *Atlas of the United States* (1835), *A New American Atlas* (various editions from 1818 to 1839), *A New Pocket Atlas of the United States* (1828), *A New Universal Atlas* (various editions 1833–44; some were published by Carey and Hart), *A New General Atlas*

(1828). Thomas R. Tanner, Stephen T. Austen, J. Knight, G. W. Boynton and J. and W. W. Warr were among engravers and cartographers employed in these works.

☆   ☆   ☆

Many maps of America have been made in Europe since the sixteenth century. So far as is known the oldest map showing America is a manuscript of 1500 by Juan de la Cosa, a pilot of Columbus. Another important manuscript map of this period is the world portolan of the Genoese Nicolay Caneiro (*circa* 1502). But the first printed map of the New World is Giovanni Matteo Contarini's fan-shaped world map of 1506, engraved by Francesco Roselli (Plate 5). Only one copy is known; it is in the British Museum.

It should be remembered that to cartographers of this date, the coast of America was thought to be the coast of Asia. Indeed in Contarini's map, it is shown as part of the Asian mainland. The West Indies are shown, but only part of South America, and that is completely divided from North America by a large ocean.[1]

The first map to name America was Martin Waldseemüller's woodblock world map of 1507. On this the New World is misshapen, but it is shown separated from Asia. The gap between North and South America remains, but it has been reduced to a mere strait. A later Waldseemüller map, the *Carta Marina* of 1516, also shows America; it is based on Caneiro's world portolan. These two maps had a wide influence on other cartographers, but out of what must have been quite numerous editions, only one of each remains. They are in the Schloss Wolfegg, Württemberg.

Other important early world maps showing America appeared in various editions of Ptolemy: that of 1508 by Johann Ruysch; the heart-shaped edition of 1511; and the edition of 1513. The 1513 edition contained a separate map of America, *Tabula Terra Nove*, in addition the world map, known as the 'Admiral's map'. Peter Apian made a copy of Waldseemüller's 1507 world map, in which the sea is still shown flowing between North and South America. The various editions of Apian's *Cosmography* between 1534 and 1584 contained world maps in which America was featured. A map of the West Indies appeared in the *Decades* of the Italian Peter Martyr or Pietro Martyr d'Anghiera (1455–1526); maps of Mexico City and some of the West Indian Islands in

[1] *See also* page 28.

the *Isolario* of Benedetto Bordone (1460–1531); and there was Oronce Finé's double heart-shaped world map of 1531, *Nova et integra Universi Orbis Descriptio*. Simon Grynaeus used Finé's map in the book *Novus Orbis Regionum* (Paris 1532); it was used also by Mercator (1538).[1]   Yet another early map that should be mentioned is Sebastian Münster's *Die Nüw Welt* (Basle 1540; reprinted 1545).   In this the outline of the New World is very distorted, though South America is more recognisable than North America.   Magellan's ship is shown on it sailing in the Pacific (Plate 58).   A Spanish map of the New World appeared in Pedro de Medina's *Arte de navegar* (Valladolid, 1545).   This was reissued in a Seville edition in 1549.

To come to the great cartographers of the sixteenth century, America was featured in the atlases of Mercator and Ortelius.   On the whole these maps, which are highly decorative, give a comparatively accurate delineation of the areas portrayed.   Maps of America appear also in the works of De Bry, the van Langrens (Plate 11), Hakluyt and Linschóten, which appeared at about this time, and in the various editions of Cornelius Wytfleet's *Descriptionis Ptolemaicae augumentum* (1597 to 1615).

In the seventeenth century maps of the New World appeared in the great atlases of Blaeu (Plate 14), De Wit, Jansson, Sanson (Plate 17) and Visscher, in addition to many others.   Not only general maps, but maps of individual states were featured, and in some cases views were added.   Blaeu's maps are magnificent in every way—engraving, drawing, calligraphy, style; and although the others sometimes approach them, they never excel them.

Maps appeared in many geographical books.   Among them may be mentioned Captain John Smith's map of Virginia (1606 and other editions) and of New England (from his *Advertisement for the Unexperienced Planters of New England*, London, 1631); William Wood's woodcut map of 'The south part of New-England' from his *New England Prospect* (London, 1634); and Johannes de Laet's *Nieuvve Wereldt* (Leyden, 1625), with ten maps by Hessel Gerritsz.

John Speed's *Prospect of the Most Famous Parts of the World* of 1627[2] contained a general map of the whole American continent, and a separate map of the Bermudas. The general map is decorated with figures of Indians and views of towns.   In an edition of the *Prospect* issued in 1676 maps were added of Virginia and Carolina, Jamaica and Barbados, New England and New York.

The eighteenth century saw an increase in the number of maps of America.   All the great cartographers issued them.   The decorative maps of Hermann Moll, issued, *circa*

[1] *See also* page 32.   [2] *See also* page 67.

1710–15, call for especial comment. They contain vignette views of local industries, scenery or native fauna. Also of outstanding interest are the twenty-one maps of Henry Popple's *A Map of the British Empire in America* (London, 1732)—a key map and twenty sections. This work was the most detailed map of America issued thus far.

John Senex issued a series of attractive and decorative maps of the New World, including some small-scale ones, in the *New General Atlas* of 1721. Thomas Jefferys's thirty-map *American Atlas* devoted, as its name implies, to the New World, appeared in 1776, and other editions followed. There were a number of other atlases devoted exclusively to the New World. Jefferys's forty-map *West Indian Atlas* appeared in 1775, and was re-issued in 1780. Jefferys also issued, in 1755, John Mitchell's famous *Map of the British and French Dominions in North America*, which depicts eastern North America from the Mississippi delta to the southern shores of Hudson Bay. In the same year he also issued a *New Map of Nova Scotia and Cape Britain* (Plate 43). The former was used in the peace negotiations between Britain and the American colonies in 1782 for the demarcation of the boundary between America and Canada.

Jefferys's successor, William Faden, issued a thirty-four map *North American Atlas* in 1777. About the same time the maps of the *American Neptune* of J. F. W. des Barres (*fl.* 1774–81) began to appear. This work was completed in four volumes, although collections vary considerably, as they were made up according to individual requirements. The charts are very detailed; they are decorative also, for many of them contain views, although some of these were issued as separate plates.

At the end of the eighteenth and beginning of the nineteenth centuries a series of large-scale maps of various parts of America was issued by Aaron Arrowsmith.

In addition to the work of the cartographers mentioned above, good maps of America, some of them very attractive, were published by Bodenehr, D'Anville, De Fer, de l'Isle, Du Val, Homann, Janvier, Köhler, Le Rouge and Seutter.

BIBLIOGRAPHY

BAGROW, LEO: *History of Cartography*. 1964.
BIGELOW, J.: 'The So-called Bartholomew Columbus Map of 1506'. *Geographical Review*. 1935.
BROWN, LLOYD A.: *The Story of Maps*. 1949.

BURLAND, C. A.: 'American Indian Map Makers'. *Geographical Magazine*. 1947.

CAMPBELL, TONY: 'The Printed Maps of Barbados'. *Map Collectors' Series*, Vol. III.

CAREY, C. H.: 'Some Early Maps and Myths'. *Oregon Historical Quarterly*. 1929.

Catalogue of Maps of the Geographic Board Ottawa. 1922.

CHAPIN, H. M.: *Check List of the Maps of Rhode Island*. 1918.

Contarini's Map of the World (British Museum). 1924.

CUMMING, W. P.: *The South-east in early Maps*. 1958.

FITE, E. D. and FREEMAN, A.: *A Book of Old Maps Delineating American History*. 1926.

FORDHAM, ANGELA: 'Falkland Islands'. *Map Collectors' Series*, Vol. II.

HARRISSE, H.: *Discovery of North America with an Essay on the Early Cartography of the New World*. 1892.

HASKELL, D. C.: *Manhattan Maps: a co-operative List*. 1931.

HOLMDEN, H. R.: *Catalogue of Maps. Plans and Charts in the Map Room of the Dominion Archives Ottawa*. 1912.

KAPP, KIT S.: 'The Printed Maps of Jamaica up to 1825'. *Map Collectors Series*, Vol. V.

KARPINSKI, L. C.: *Bibliography of the Printed Maps of Michigan 1804–1880*. 1931.

LE GEAR, C. E.: *United States Atlases . . . in the Library of Congress*. Washington. 1950.

List of Maps of Boston published between 1600–1903. Boston. 1903.

LOWERY, W.: *A Descriptive List of Maps in the Spanish Possessions within the present Limits of the United States 1502–1820*. 1912.

Map Collection of the Public Reference Library Toronto. 1923.

NUNN, G. E.: *The Mappemonde of Juan de la Cosa*. 1935.

PALMER, MARGARET: 'The Printed Maps of Bermuda'. *Map Collectors' Series*, Vol. II.

PHILLIPS, P. LEE: *Alaska and the N.W. Part of North America 1588–1898*. Washington. 1898.

—— *List of Maps and Views of Washington and District of Columbia in Library of Congress*. Washington. 1900.

RAISZ, E.: 'Outline of the History of American Cartography'. *Isis*, Vol. XXVI. 1937.

SHILSTONE, E. M.: 'List of Maps of Barbados'. *Barbados Museum and Historical Society Journal*. 1938.

SKELTON, R. A. and TOOLEY, R. V.: 'The Marine Surveys of James Cook in North America 1758–1768'. *Map Collectors' Series*, Vol. IV.

SOULSBY, B. H.: 'The First Map containing the name America'. *Geographical Journal*. 1902.

STEVENS, HENRY: *Notes Biographical and Bibliographical on the Atlantic Neptune*. 1937.

STEVENSON, E. L.: *Early Spanish Cartography in the New World*. 1909.

STOKES, I. N. PHELPS: *The Iconography of Manhattan Island*. Vol. II. 1916.

THOMPSON, E.: *Maps of Connecticut before 1800*. 1940.

TOOLEY, R. V.: 'California as an Island'. *Map Collectors' Series*, Vol. I.

—— *Maps and map-makers*, 1962.

H

VERNER, COOLIE: 'Mr. Jefferson makes a Map'.   *Imago Mundi*, Vol. XIV, pp. 96–108.   1959.

WAGNER, H. R.: *The Cartography of the Northwest Coast of America to the Year 1800.*   2 vols.   1937.

WHEAT, C. I.: *Mapping the Trans-Mississippi West.*   5 vols.   1957–63.

WROTH, L. C.: 'The early Cartography of the Pacific'.   *Papers of the Bibliographical Society of America*, Vol. XXXVIII.   1944.

CHAPTER TEN

Australasia

CENTURIES before Australia was discovered—even as far back as Greek times—man thought that there was a great southern continent. Mercator shared this belief for in his world map of 1569 he showed three great land masses: Eurasia and Africa, North and South America (or the New Indies), and *Continens australis*, or the southern continent. The shape of this continent as he showed it is much bigger and more sprawling than the reality. Tierra del Fuego, for example, is incorporated in it, and from there it stretched westwards and eastwards to encircle the whole southern polar area. It is at least possible that the coastline as shown near New Guinea shows some remote knowledge of the Australian coastline. Marco Polo had mentioned lands far to the south of China, and possibly some old Chinese navigator or traveller had seen or heard something of Australia, and so helped to implant the legend in the folklore of his country. Other cartographers besides Mercator, Gastaldi and Ortelius among them, showed the apocryphal southern continent in much the same form as that in which he drew it.

It was long before Australia received its present name. The Dutch for many years called it Company's New Netherlands, later New Holland. But there had been several other names for it, including Terra Australis Incognita, Land of the South, Land of Eendracht, Land of Beach, Patalie Regio, Ultimaroa, Carpentaria. Australia became the generally accepted name during the nineteenth century. Tasmania, named after its discoverer, Abel Janszoon Tasman (1603–59), was for long known as Van Diemen's Land, after Anthony Van Diemen (d. 1645), the Dutch explorer and colonial governor who sent Tasman on the voyage during which it was discovered. Tasman so named it in his superior's honour, but British colonists later re-named it Tasmania. The name of New Zealand also recalls the influence of the Dutch navigators in the area.

Although Australia was first discovered by the Dutch in 1605, no separate map of it

was attempted until the nineteenth century, although it was regularly featured, sometimes distorted, and sometimes incomplete, on world maps. Tasmania and New
Guinea—even New Zealand—were sometimes shown joined to Australia, as in the
world maps of Du Val (1676), Nolin (1705), de l'Isle (1714 and 1730: Plate 18), Buache
(1741), Vaugondy (1752), Le Rouge (1778) and many others. Little or nothing was
known of its interior, and even when its coastlines were depicted with comparative
accuracy, no inland features were shown.

Of early navigators in Australasia, Tasman made the greatest contribution to its
cartography, collecting and collating the work of previous explorers, and adding his
own discoveries, in particular those of Tasmania and parts of New Zealand. Captain
James Cook took all of this a stage further in the late 1760's and early 1770's, by
completing the outline of the coast of New Zealand and the eastern part of Australia.
Other navigators completed the work by the end of the first quarter of the nineteenth
century. The interior remained unmapped altogether until the nineteenth century,
and much of it is still unsurveyed.

During the whole of this long period, from the seventeenth to the latter part of the
nineteenth century, maps of Australia and the south-western Pacific were issued by
greater and lesser cartographers, on which the navigators' new discoveries were shown.
Among them may be mentioned Blaeu's 1640 map of the East Indies, with part of
Carpentaria[1] shown; Hondius's world map of *circa* 1590 (Plate 13); the *Carta particolare
della costa Australe scoperta dall' Olandesi* of Sir Robert Dudley (1593–1649); Jansson's
1652 map of the East Indies with Australia shown as 'Terra del Zur'; Alphen's rare map
of the East Indies (1660) from his *Zee Atlas*, which shows Tasman's discoveries, New
Zealand and Tasmania: and Thevenot's map of Australasia from his *Relation de divers
voyages*. This last was copied from Blaeu's map of the East Indies and was re-issued by
Bowen in 1747. Maps were also published by Allard (1676), Danckerts (1680),
Mortier (1700), Roggeveen (1676) and Van Keulen (1680), all based on Tasman's discoveries. Later world maps incorporating new information concerning Australasia
from Cook's *Voyages* (1773–84), were those by Faden (1783, 1786, 1790), Kitchin
(1787), Bowles (1790), Laurie and Whittle (1794, 1799), Cary (1801, 1811, 1819) and
Arrowsmith (1834).

Late inaccuracies in the representation of Australia are found in maps showing
Tasmania joined to the mainland, and in those which gave an incorrect outline to the

---

[1] The Gulf of Carpentaria is on the north coast of Australia between Capes York and Arnhem. It was named
in 1623 after Pieter Carpentier, Governor-Gerneral of the Dutch Indies.

eastern part of the south coast of the mainland. Examples of the first may be seen in maps by Bowen (1760), Denis (1764), Faden (1787) and Laurie and Whittle (1797); and of the second in those of Bowen (1777), Kitchin (1787), Laurie and Whittle (1794 and 1797) and Cary (1801, 1806, 1811 and 1819).

Salient inland features such as mountains began to appear on maps at the end of the eighteenth century. Rivers appeared a little later, between 1800 and 1810. More detailed local maps began to be published in 1810. These included Booth's *Plan of the Settlements of Australia* (1810), Mitchell's map of New South Wales (1834), Evans's chart of Van Diemen's Land (1822) and Captain Smith's *Plan of the Town of Wellington* (1840). Maps of the Australian states also appeared in such general atlases as those of L. Tallis (1851), which were decorated with views and figures, and J. Arrowsmith (1832). Among Australian cartographers of this period was Thomas Ham, who also printed the famous 'half-length' and 'Queen enthroned' stamps of Victoria.

BIBLIOGRAPHY

CALVERT, A. F.: *The Discovery of Australia.* 1893. 1902.

COLLINGRIDGE, G.: *The Discovery of Australia.* 1895.

CRONE, G. R.: 'The Discovery of Tasmania and New Zealand'. *Geographical Journal.* 1948.

HARGREAVES, R. P.: 'French Explorers' Maps of New Zealand'. Vol. III. *Map Collectors' Series.*

HOCKEN, T. M.: 'Some Account of the earliest Maps relating to New Zealand'. *Transactions and Proceedings, New Zealand Institution.* 1894.

MACFADDEN, C. H.: *A Bibliography of Pacific Area Maps.* 1941.

QUIRINO, C.: *Philippine Cartography 1320–1899.* 1959.

TOOLEY, R. V.: 'Early Maps of Australia, the Dutch Period'. Vol. III, *Map Collectors' Series.*
—— *Maps and Map-Makers.* 1962.
—— 'One Hundred Foreign Maps of Australia'. Vol. II, *Map Collectors' Series.*
—— 'Printed Maps of New South Wales'. Vol. V, *Map Collectors' Series.*
—— 'The Printed Maps of Tasmania'. Vol. I, *Map Collectors' Series.*

WROTH, L. C.: *The early Cartography of the Pacific.* 1944.

General Bibliography

BAGROW, LEO: *History of Cartography*. 1964.

BEAZLEY, C. R.: *The Dawn of Modern Geography*. 3 vols. 1897–1906.

BROWN, L. A.: *Mapmaking: The Art that became a Science*. 1960.

—— *The Story of Maps*. 1949.

—— *The World Encompassed*. An Exhibition of the History of Maps held at the Baltimore Museum of Art. 1952.

COLLINDER, PER: *History of Marine Navigation*. 1954.

CRONE, G. R.: *Maps and their Makers*. 1968 and various earlier editions. The 1968 edition is enlarged and its bibliography expanded.

CURNOW, I. J.: *The World mapped*. 1930.

DICKINSON, R. E. and HOWARTH, O. J. R.: *The Making of Geography*. 1933.

FORDHAM, SIR HERBERT GEORGE: *Maps, their History, Characteristics and Uses*. 1927.

—— *Some notable Surveyors and Map-makers of the 16th, 17th and 18th centuries and their work*. 1929.

HINKS, A. R.: *Maps and Survey*. 1944.

JERVIS, W. W.: *The World in Maps*. 1936.

KOEMAN, C.: *Collections of Maps and Atlases in the Netherlands: their History and present State*. 1961.

LISTER, RAYMOND: *How to Identify Old Maps and Globes*. 1965.

MAGGS, F. B.: *Voyages and Travels in all Parts of the World*. 5 vols. 1942–63.

MAP COLLECTORS' CIRCLE: A series of specialised publications issued to subscribers. Address: Durrant House, Chiswell Street, LONDON, E.C.1.

PHILLIPS, PHILIP LEE: *A List of Geographical Atlases in the Library of Congress*. 5 vols. 1909–58.

RADFORD, P. J.: *Antique Maps*. 1965.

RAISZ, E.: *General Cartography*. 1948.

REEVES, E. A.: *Maps and Map-making*. 1910.

RISTOW, W. W. and LE GEAR, C. E.: *A Guide to Historical Cartography*. 1960.

SKELTON, R. A.: *Decorative printed Maps of the 15th to 18th Centuries*. 1952.

SKELTON, R. A.: *Explorers' Maps*.   1958.

STEVENSON, E. L.: *Portolan Charts: their Origin and Characteristics*.   1911.

TAYLOR, E. G. R.: *The Haven-finding Art*.   1956.

—— 'The South-pointing needle'.   *Imago Mundi*, Vol. VIII.   1951.

TOOLEY, R. V.: *Maps and map-makers*.   1962.

TOOLEY, R. V., BRICKER, C. and CRONE, G. R.: *A History of Cartography*.   1969.

WINTER, HEINRICH: 'A Late Portolan Chart at Madrid and Late Portolan Charts in general'.
    Imago Mundi, Vol. VII.   1950.

WINTERBOTHAM, H. S. L.: *A Key to Maps*.   1936.

Index